*Inside your head, Wit.......... ....
pavement beneath your feet, and the trees in the distance.
It lingers in the tales told by old women, in folk traditions
and long forgotten rituals. It is as real as you and I, and
I'm inviting you to walk there.*

Take my hand, go to https://enter.witheredhill.place

Praise for *Withered Hill*

'Eerie, erotic and engrossing, *Withered Hill* is a masterpiece of folkloric horror, one I suspect shall henceforth be mentioned in the same breath as *The Wicker Man*, as it well and truly deserves to be' **Kealan Patrick Burke**

'A folk horror novel that clamps tighter with every page. Barnett's writing is aces and *Withered Hill* is the ultimate nightmare destination' **Laird Barron**

'Gets into you as the best folk horror tends to: the woods grow ever-deeper, the villagers more smilingly hostile as the escape routes shrink tighter than a keyhole. But there's much delight in being trapped in the hands of such an agile and compelling storyteller' **Nick Cutter**

'A bloody masterpiece! I tore through it. The glib description is *Black Mirror* meets *The Wicker Man*, but it's so much more than that. I loved every word!' **Mark Stay**

'*Midsommar*, *Evil Dead* and *Stepford Wives* all in one, this twisted dark fairytale rooted in the modern world had me too scared to walk my dog in broad daylight. Masterful storytelling, sinister, sick and funny – if you're as warped as this folklore horror gem' **Lisa Rookes**

'Beguiling, bewitching and brilliant. *Withered Hill* gets under your skin like some long-forgotten pagan spell and keeps you turning the pages late into the night' **Jeremy Dyson**

DAVID BARNETT

WITHERED HILL

CANELO

First published in the United Kingdom in 2024 by

Canelo
Unit 9, 5th Floor
Cargo Works, 1-2 Hatfields
London SE1 9PG
United Kingdom

Print ISBN 978 1 80436 751 3
Ebook ISBN 978 1 80436 757 5

Cover design by Cover design by Sarah Whittaker

Look for more great books at www.canelo.co

Printed and bound in Great Britain by Clays Ltd, Elcograf S.p.A.

1

To Alice. A warning to be good...

As I were going up Withered Hill
With night-time coming soon
I met a man under the trees
Whiter than the moon
He smiled at me and stroked my hair
I were frit for my life
He showed his teeth and smiled and said,
'Owd Hob wants a wife'

Pennine Lancashire playground rhyme,
circa nineteenth century

1

Outside

Days to Withered Hill: 30

Everyone leaves eventually.

Jill had got married and Liz had got pregnant, quietly removing themselves from Sophie's orbit and beginning to circle the more important things in their lives. And now Donna was joining them.

Sophie was drunk. In fact, she'd gone past drunk, far past it. The shots weren't touching the sides, her drinks disappearing before anyone else's, resulting in her groping around for other glasses on the table, finishing off the dregs. She leaned heavily on Donna, sitting on the stool beside her.

'I don't want you to go to Dubai.'

'Sophie!' admonished Liz. She was checking her phone every ten minutes, no doubt wondering if it was OK to call the babysitter yet again, Sophie thought. 'You're supposed to be happy for Donna!'

'I am happy for her,' said Sophie, which she and everyone else knew was a lie. Sophie felt like she might start crying. Had that already happened once tonight? She wasn't entirely sure.

Donna put her arm around her and hugged her close. 'I'm not going to the moon. I'll be home for visits, and you can come out and see me.'

Sophie thought about the pile of unopened envelopes on her table, most of which would be stamped *overdue*. Yeah, she was really going to be jetting off to Dubai, the playground of the rich and famous, for a weekend.

'You know what you need,' said Jill. Her finger tapped her glass, the still-shining wedding ring beating out the rhythm of the song that was playing in the bar.

Sophie covered her face with her hands. 'Please don't say a man,' she said, her voice muffled. She *did* need a man; at least, she wanted one. But not a bastard. And they seemed to be the only sort she ever met. She needed money. She needed a holiday. She needed a dozen things.

'I was going to say a job,' said Jill.

Liz stood up and announced, 'I'll get us another round of drinks. Same again?'

'And more shots,' said Sophie. She turned back to Jill. 'I've got a job.'

Jill leaned her hand across the table onto Sophie's. Sophie felt the cold weight of her engagement and wedding rings on the backs of her fingers. Jill didn't have a job. Not any more. Not since she'd married Pete. So she had no place to talk. Sophie could feel those very words leaving her mouth, then bit down on them.

'Temping isn't a job,' said Jill. 'Not a proper one. Not for a thirty-two-year-old. You're smart, Sophie. Intelligent. Clever. But you're drifting. That's OK when you're in your twenties, but... surely there's something you can do? Something to make you happier?'

Sophie stood up, abruptly, her legs knocking the table, shaking the growing forest of empty glasses. 'What would make me happier is if people would just stop *leaving*,' she said unsteadily. 'I need a bit of fresh air.'

Outside the pub, she leaned on the wall, eyes closed, listening to the sounds of the London night. People just kept slipping through her fingers. Nobody cared about how she felt.

When she opened her eyes again, there was a man, a little way down the outside of the pub, leaning on the wall as well, staring at her. He was shorter than Sophie, with straight hair that looked like his mother had cut it with a pudding bowl over his head. His fringe was jagged and hung over his right eye. He was about her age, she figured, but not the sort of man she'd ever go for in a million years. His trousers were showing too much sock, and not in a hipstery kind of way, but because he had them belted up too high, a crisp blue checked shirt tucked into them.

'I'm Colin,' he said.

The headlights of a car turning past the pub swept over them, and she saw that underneath his jagged fringe he had a lazy eye.

Sophie didn't offer her name. She looked him up and down appraisingly. 'I'm not going to sleep with you.'

'I wasn't going to ask you to,' said Colin. He dug into his trouser pocket and pulled out a business card.

'I don't want your number, either.'

'It's not my number. It's my boss's. Mandy.'

Sophie took the card and stared at it through blurry eyes. 'Why do I want your boss's number?'

'Because she's recruiting. I overheard you talking to your friends. About a job. It's good money. Easy work, if a little boring. Call her tomorrow.'

3

Sophie laughed. 'Yeah, I bet. Pimp is she, this Mandy? Anyway, tomorrow's Saturday.' She screwed the card up in her fist and threw it on the floor.

'Nothing like that.' He seemed jumpy, bending down and retrieving the card, smoothing it out in his hands. 'Data entry. And Mandy works twenty-four-seven. Anyway, sorry to have bothered you.'

He walked back towards the pub, brushing past her. Sophie watched him go through one eye, and was about to go back in when a man came out and stopped in the doorway, lighting a cigarette, his hand cupping the flame from his lighter and illuminating a face that was, she decided, far more her type.

'Have you got a light?' asked Sophie.

He shrugged, and gave her a good long look, then handed over the Zippo.

'Have you got a cigarette?' she asked. She didn't even smoke. But she did know an opportunity when she saw it.

–

The buzzing of a phone woke Sophie up. Her mouth was as dry as the Sahara and when she lifted her head off the pillow, it began to thump. It wasn't her phone, though she saw she had a dozen missed calls when she looked at it through one eye.

He was standing by the door to her bedroom, pulling up his jeans, his phone cradled between neck and jaw. 'Yeah, yeah, I'm on my way,' he said. Sophie couldn't even remember his name.

He glanced at her and forced a tight smile. He wasn't quite as handsome as she'd thought he was the night before.

4

She glanced over at the other side of the bed and saw the open condom wrapper on the bedside table. She couldn't remember the sex, had no idea if it was any good.

'Got to go,' he said. And he went.

Everyone left… eventually. Some sooner rather than later.

Sophie suddenly felt embarrassed by her nakedness under the duvet. She inspected her phone as the door slammed. There was a message from Jill.

> Hope you're OK. Tried to call you after you disappeared. Still up for brunch before Donna's plane leaves?

Brunch was at a country pub not far from Heathrow. It took Sophie a Tube, train and taxi to get to it. Everything went on her credit card. She sent up a silent prayer each time she tapped, hoping it would hold out.

The others were already there, Liz with her baby bouncing on her knee. Sophie made a show of fawning over it, though babies made her feel… weird. After what had happened to Emily, she supposed. Everyone knew about that, of course, but it was a long time ago. Before any of them knew Sophie. She had been only nine herself.

'My treat,' said Donna as they ordered brunch and a couple of bottles of Prosecco.

Sophie half-heartedly joined in with the protestations, but Donna was firm on it. She just had to get through the rest of the weekend. Payday from the last job was on Monday.

'So, what happened to you?' asked Donna, pouring the fizz. 'Or should I say, *who*?'

'I wish you wouldn't go off like that,' said Jill reprovingly. 'Without a word. Anything could have happened to you.'

'I bet anything did,' remarked Liz with a snort.

Sophie sipped her Prosecco and thought about the way that guy had looked at her when he'd left her flat. Dead-eyed, disinterested, a little ashamed in the cold light of day. She wondered who was calling him? His girlfriend? Wife?

Sophie changed the subject and asked Donna to show them the photos of her new apartment again.

Too soon, the brunch was over. Jill had brought Liz in her car and was running Donna to the airport.

'Lift back to town?' she asked Sophie. 'Aren't you at the shop today?'

She had passed it not long after moving to London. It was called Little Angels, and was raising money for a charity that helped families dealing with the loss of a child. Sophie helped out every other weekend, just for a few hours on a Saturday or Sunday. It was, she thought, the least that she could do.

'I said I'd do the afternoon tomorrow,' replied Sophie. She didn't want to go to the airport and hug Donna and cry in the departures lounge. She glanced up at the clear sky. 'I'm fine. I'll stay for a bit.'

Jill nudged Liz. 'Probably meeting the guy from last night, am I right?'

Sophie smiled, and then hugged Donna tight.

When they'd all driven off, she sat there on her own for a while, finishing the Prosecco.

She was digging in her bag for her sunglasses when her fingers closed around a crumpled piece of card. She smoothed it on the table. The business card, that weird guy – Colin? – had given her last night. She could have sworn she'd thrown it away. There was a name, Amanda Scott, a mobile number, and a simple logo that said *Gemini Data Processing*.

Good pay, he'd said. Easy work. Fine to call on a Saturday. Sophie tapped her teeth with her fingernail and then quickly dialled the number.

'Mandy Scott.'

'Hello, my name is Sophie Wickham. I met, uh…' She searched for the name in her fogged memory.

'Colin?' said Mandy.

'Yes, and he said—'

'Are you free Monday? For an interview?'

Sophie blinked, and took the phone away from her ear, staring at it. 'Yes, I am, all day.'

'Let's say three o'clock,' said Mandy. 'I'll text you the address. See you then.'

Sophie stared at the phone. Mandy hadn't waited for her to say yes or no before hanging up. There was a waiter hovering at the table. 'Anything else, madam?' he asked.

Sophie took a deep breath and, under the table, checked the banking app on her phone. She breathed a sigh of relief and said, 'Yes, another bottle of that Prosecco, please.' She could just about manage it.

She sat in the sun and drank it slowly, eking it out over the next hour, until she saw what she was sure was Donna's plane, climbing steeply into the blue and banking to the east.

2

Inside

Days in Withered Hill: 357

It is only when Sophie has resigned herself to never leaving Withered Hill that they tell her she can go.

Of course, it isn't going to be quite as easy as that.

'You'll have to make a bower,' says Noah Jones, but in the thick Lancashire accent that had sounded so alien to Sophie when she first arrived, and which it has taken her a year to finally hear without having to internally translate. It sounds more like, *tha'll've'fot'mek'a'bowwer.*

'What do you mean, a bower?'

Noah chews his tobacco and spits a brown stream of it on the cobbled street outside the Post Office. It was here that Sophie had gone when she had stumbled into Withered Hill a year ago, naked, adorned in earth and twigs and leaves and insects, shorn of memory. The Post Office rose in the murk of her mind as something official, something safe. Somewhere she could get help.

'That, you'll have to find out for yourself,' says Noah.

'Is that part of the test?'

Noah looks down the cobbled hill, to where the village policeman, Constable Parry, is wheeling his ancient push-bike up in the gutter. Constable Parry had been the person she had run to, the authority figure her panicked mind vaguely recognised, on that first day.

'Aye, if you like,' says Noah, nodding. 'All part of the test. Cheerio.'

Sophie watches him walk down the street, stopping to talk to Constable Parry. It's a bright, clear day, the blue sky arching overhead until it touches the green hills and the moorland, peppered with purple heather that spreads out beyond the ring of trees surrounding the village.

A bower. What's a bower? She has a fancy that it has something to do with the woods. She's going to have to look it up.

Out of the corner of her eye, she catches a movement from within the Post Office.

It's Carol, waving at her through the window and mouthing, 'Coo-ee!'

Sophie smiles at the sturdy, grey-haired woman who hops around behind Sophie's own reflection in the window that's pocked with postcards advertising things for sale, the services of cleaning ladies, offers of vans to help with moving house, notices about galas, fetes, car-boot sales long since held. On the wide, dusty sill inside the shop, a line of dead, desiccated flies are laid out as though for some ritual.

In the cool of the shop, Carol moves from behind the plastic screen of the Post Office desk to the open general shop counter, on which a plastic half-globe sprouts flavoured lollipops like a hedgehog, next to the copies of that morning's local paper. On the front page is a picture of children enjoying themselves at a summer fun day, beneath which a headline sits uncomfortably about a paedophile being jailed. Not news from Withered Hill, of course; from the bigger villages and towns further along the valley. There is never any news from Withered Hill in the papers.

'Carol,' says Sophie, 'do you know what a bower is?'

Carol clasps her thick, gnarled fingers together at her ample breast. 'Oh! They've told you that you can go, then?' *Thiv'towd'thi'tha'can'go?* 'How lovely. I'll miss you. You've been a breath of fresh air in Withered Hill.'

'I have to make a bower.'

Carol nods enthusiastically. 'That's right.'

'I don't know what that is, though.'

Carol smiles broadly. 'I'm sure it'll all come right in the end.'

There's a huffing and puffing and stamping of feet behind her. Sophie doesn't have to turn round to see they belong to craggy-faced old Mr Winterbottom, still in his big coat despite the early-summer heat softening the horizon with smoky haze.

'Zachary, our Sophie has been told to make a bower!' announces Carol happily.

Mr Winterbottom shuffles up to the counter, coming barely up to Sophie's shoulder, hunched over in his coat that smells of sheep and swill and stale sweat. 'Aye,' he says to Carol. 'I am on the bloody parish council, you know. Two ounce of my baccy.' He finally looks at Sophie. 'Aye, be sorry to see her go.'

When he's gone, Sophie shivers slightly in the cold air the big fan behind the counter is blowing around the Post Office.

Carol smiles broadly. 'That was positively effusive for Zachary.' She claps her hands together. 'Now! You'd better be off and thinking about making your bower! So lovely!'

Sophie walks from the treeline at the north end of Withered Hill, picking twigs from her hair as she puts one uncertain step in front of the other, wiping sap from her eyes, brushing soil from her breasts. She is naked, but she doesn't know what that means. She doesn't know what anything means. She was not there, and then she was. Or so it feels like.

The open meadow reveals a path that becomes a track and eventually a narrow road. She rolls the words around her mind. Path. Track. Road. They are not new words, she knows what they mean when she summons them, but they seem somehow new to her. She tries to voice them, but her mouth is dry and full of dirt and leaves. Spitting them out, she falls to her hands and knees and spews a stream of viscous black vomit onto a rock.

The first place she comes to is what she later learns is Nut Nan Farm. It is bordered by a drystone wall and beyond a field she sees a house, made from grey slabs of stone. In the yard in front of the house are two tractors. A man, wearing shapeless brown trousers and a checked shirt with the sleeves rolled up to his elbows, displaying what she can see even from this distance is a patchwork of dark tattoos, is working on one of them, and there is a small child throwing a ball for a dog beside him. She hears a shout from the child and stops to watch from the road. The child, a girl, is pointing at her and the man straightens up, shielding his eyes from the sun that burns directly overhead, to stare at her.

She has the first inkling, the first notion, that she is naked, then. She thinks she should be covered. She looks down at her body, slim and streaked with mud and scratches. Her feet are caked with dirt.

When she looks up again, the man is sending the child running into the farmhouse, the black and white dog yapping at her heels. She comes out a moment later with a broad woman in a green dress, wiping her hands on a tea towel. The small girl is dragging a sack with her.

Help me. The words form in Sophie's mind. She tries to say them, but they come out as a dry, ashen croak.

Water. That's what she needs. Water. To drink and to cleanse herself. Perhaps there is water in the sack. Perhaps they are going to help her even though she cannot voice the words.

But it is not water. The man pulls out a... mask? No, a pig's head. Hollowed out, pink and leathery. He puts it on top of his own head, as though it is a hat, then pushes it down until it covers his face. His wife does the same with another from the bag. Then they both help the child with her own.

The three of them stand and stare at her, a family of pig-faced people. Sophie stares back. Then they all lift their right arms and point further along the road which turns upwards in a gradual slope, and where nestles on the hillside the village.

Sophie fixes her eyes on the cluster of roofs and sets off.

A little further on the road bordered on both sides by stone walls, the sun beating down upon her shoulders and back, Sophie sees a red van. Something rises in her fogged mind. This is... official. This is something to be trusted. There is a figure sitting in the van, which is pointing towards her. She pauses and looks back, beyond the farm. The road... where does it go? Should she be walking that way, not this? But this is where the pig-family told her to come.

A post van. The mail. That's what it is. Letters sent by people to other people in other places. The thought makes her frown. Other places. What other places?

As she draws closer to the red van, the window at the side begins to slide down.

Inside is a man wearing a pale blue shirt and dark shorts that come to his knees. He has the face of a rabbit. The mask covers his entire head, the ears folding against the roof of the van cab. It is made of grey fur and the whiskers are wires. His eyes are blue behind the holes. He leans a little way out of the window and twists around, pointing back towards the village.

'Water,' Sophie manages to say. 'Water…'

Beside him, on the empty seat, she can see a plastic bottle. He follows her gaze and his eyes narrow in a frown. Shaking his head, he points along the road again. Then he winds up the window.

–

The tarmac road abruptly becomes a matrix of cobbles as she enters the village, rows of small, neat cottages on either side. It rises steeply now, and Sophie feels sharp stones pricking at the soles of her feet. There's a movement to her left, a swishing of a curtain, then an identical one to her right. At the end of the row, on the left, is a pub with a thatched roof, a sign creaking in the breeze. In flaking paint, it depicts a man in a smock with a scythe and a sheaf of wheat, and peeking from behind him is a red, lascivious figure with goat-like horns and a rattish tail. Painted on the whitewashed wall of the pub in green letters are the words *The Farmer and Devil*.

There are three men sitting at a wooden table outside the pub, young and strong and athletic, their bare forearms tanned, their jeans tight. They wear the masks of black dogs, their noses glistening as they watch her approach. Their heads nod up and down as they look at her, naked and filthy, in front of them. They all point up the hill.

Outside a butcher's shop, a stout man in a red and white striped apron stands, already pointing as she passes. He wears a calf's head, a real one, its eyes milky and glassy, flies swarming over it, blood running from its ragged, severed neck and staining his white tunic.

Something rises inside of Sophie. A bubble of a sob, encasing fear and confusion and the growing and unbearable need to run, to flee, to be away from this horrible place. Ahead of her, she sees a sign. *Post Office*. Letters. To other places. *Other* places.

She half-runs, half-staggers up the hill, clinging on to some half-submerged kernel of knowledge. This is official. This is safe. This is where she can get help.

Sophie bursts through the door, the heat of the sun giving way to a dry, dusty coolness presided over by a woman who Sophie will come to know as Carol. She is not wearing a mask. Sophie feels like crying.

The woman clasps her hands together and says, 'Oh! You're here! Lovely! Welcome to Withered Hill!'

3

Inside

Days in Withered Hill: 357

It is almost noon, so Sophie decides to go over to St Michael's School and see if Catherine would like to take lunch with her. She gets to the school, behind The Farmer and Devil, a little early and Catherine is still teaching her class of nine- and ten-year-olds; she sees Sophie through the window and waves her in.

Catherine is her best friend in Withered Hill. Her confidante. If anyone in the village is going to give her any idea of what she has to do about this bower question, it will be Catherine.

Sophie lets herself into the school building and Mrs Mangnall, the receptionist and secretary, smiles at her from behind her desk. 'I heard they'd asked you to make a bower, dear Sophie,' says the sparrow-thin old lady with a pair of glasses hanging at her slim chest on a chain around her neck. 'I'm so pleased for you!'

Quietly, Sophie sneaks into the classroom where Catherine is teaching, raising a hand in apology as twelve faces turn to look at her. There is a murmur of whispered voices.

Catherine raps her desk with a wooden ruler. 'Look at me, please! The lesson is still in progress!'

Where Sophie is tall, dark-haired and pale-skinned, Catherine is short, blonde and tanned. Good Lancashire farming stock. She is very beautiful, with piercing green eyes and full lips. Half of the men in the village had tried to woo her at the Faunalia festival, their collective ardour melting the February snow that had blanketed Withered Hill and the surrounding moors for weeks. Young men, old men, bachelors, married, it didn't matter. They were like dogs following her around, as though she exuded a scent that robbed men of their senses. Catherine had not – as far as Sophie knows, at least – capitulated to any of their advances. She feels a pang in her heart as she remembers that Faunalia night. She will miss Catherine terribly when she finally gets to leave.

On the blackboard behind Catherine is a carefully drawn and detailed chalk map showing Withered Hill, surrounded by its ring of trees, and beyond those to the south, three coloured rectangular blocks that, Sophie knows, represent the Parliamentarian forces advancing on what they believed to be a Royalist stronghold. Their actual, eventual target was Preston, and in August of 1648 they – or what was left of them – would be part of the Cromwellian force that fought a battle there that heralded the end of the Second English Civil War. Withered Hill was merely a chance encounter as they pushed towards the defining skirmish.

Catherine is wearing a tight black pencil skirt and a sleeveless, low-cut top that accentuates her curves. Catherine was her first true friend in Withered Hill, and Sophie feels another stab of impending loss just looking at her. She brushes it away. Since she arrived in Withered Hill, she has been trying to leave. Now she can. She has friends outside; she has family. She will forge those relationships anew and make fresh ones. She might even meet someone just like Catherine…

She knows she won't.

'So the Parliamentarian forces are gathered here, here, and here,' says Catherine, tapping the end of her ruler on each of the coloured blocks.

A boy near the back of the class shouts, 'Boo!'

The others giggle and Catherine quiets them with a stern stare. 'They were not our enemies, in the sense that Withered Hill was not a Royalist enclave as they suspected, though it was indeed an enclave of a sort they were rather *not* expecting.' she says, casting her gaze around the class. 'Lucy, what does *enclave* mean?'

A girl with red hair in pigtails says hesitantly, 'Is it… like a place where people different from everyone else around them live, Miss Baldwin?'

'Very good, Lucy,' says Catherine, nodding. 'They had sent in a small envoy to instruct Withered Hill to surrender or be burned to the ground the following morning, not listening to the protestations that the village supported neither Cromwell nor the King.' She smiles wryly, and glances at Sophie. 'They thought you had to be either one thing or the other, which is always the case with the outside world. Now, who can tell me what happened as the sun rose over Withered Hill that summer morning in 1648?'

'The soldiers were all dead, miss,' says a crop-headed boy with a thick lisp.

'Not quite all, Samuel, but a good number of them. Enough that the forces hurriedly packed up and departed, and left Withered Hill well alone, preferring to take on a battle they had more chance of winning. And who had saved Withered Hill that night from the marauding Cromwellian forces?'

'Owd Hob!' say the children in unison in sing-song voices.

Catherine smiles, just as a shrill bell sounds in the corner of the room. 'Yes. Indeed. Owd Hob. Now, off to lunch, the lot of you. And back here when the next bell sounds. We have maths all afternoon.'

The children groan collectively and drag themselves out of their chairs. They file past Sophie out of the classroom, some of them stealing shy glances at her. When the last one has shut the door behind him, Catherine lets out a huge sigh.

'What a morning.' She roots in her black handbag for a packet of cigarettes and a lighter. 'Have you eaten?'

Sophie shakes her head.

Catherine says, 'Let's go over to the pub, then. I'll stand you a ploughman's in the beer garden. I believe you have news?'

Sophie sticks out her bottom lip. 'Did everyone know before me?'

Catherine laughs. 'Of course. This is Withered Hill. Come on, I'm dying for a ciggie.'

Days in Withered Hill: 1

Carol closes up the Post Office and leads Sophie up a narrow, dark flight of stairs to a flat above, where she makes a pot of strong tea in the kitchenette and pours her a cup, adding sugar and milk. Tea. Sophie sips at it, feels it scalding her throat and the inside of her belly. Her empty belly. She thinks she might throw up again, but she keeps sipping at the tea and the feeling goes away. In the next room, she hears a gushing of water as Carol runs a bath.

'Are you hungry, dear?' asks Carol, wiping her hands on her apron. 'I've got a lovely lemon drizzle cake.'

'Hungry,' says Sophie, trying out the word.

Carol sits beside her and puts a hand on her filthy thigh. 'Oh, you poor thing, you're all at sixes and sevens, aren't you? Do you know your name?'

Sophie thinks about this for a moment. 'Sophie... Sophie... something.'

Carol pats her thigh. 'It'll come to you. We'll have you up to speed in no time.' She cocks her head on one side. 'I think we'll get you cleaned up first, then some cake.'

In the bath, the water quickly turns black as Carol sets about scrubbing her with a rough, hard sponge. Sophie closes her eyes and holds her nose as Carol pours a jug of water over her head, shampoos her hair, and then does it again.

'Clean as a whistle!' she declares. 'Up you get.'

Sophie stands up, her ankles disappearing beneath the black water, and looks down at her clean body. Her dark hair is twisted in a rope and falling over one shoulder, covering her breast.

'Lovely body on you,' observes Carol matter-of-factly as she wraps a towel around Sophie's shoulders. 'Nice childbearing hips. You'll have all the boys chasing you come Beltane.'

'Where am I?' says Sophie as Carol takes her hand and helps her to step out of the bath onto a mat.

'I already told you, silly,' says Carol with mock sternness. 'Withered Hill. You're in Withered Hill. Where you belong.'

'Withered Hill,' repeats Sophie. 'Withered Hill. But where is that?'

Carol shoots her a reproving look, as though she's being either very silly or very dim. 'It's home, you daft apeth. You came here, didn't you? You came home.'

She feels Carol briskly drying her between her legs with a towel. 'I think I might like that cake, now.'

'Lovely!' Carol beams. 'I'll get you some clothes. We can have some more tea and cake and a lovely chat. Then I'll take you to see the others.'

'Others?' says Sophie, frowning. 'Those people with the… the heads? The masks?'

Carol waves her hand. 'Oh, don't mind them. That's just one of our silly little traditions here. You'll get used to that sort of thing. There's plenty who want to see you, my girl. They're all dying to meet you. To welcome you home.'

'Home,' says Sophie as Carol kneels down in front of her and instructs her to lift one foot, then the other, and pulls up a pair of knickers.

She pats Sophie's bum. 'Oh, yes, they're all dying to meet you. And the boys are going to go crackers for you, young lady. Catherine Baldwin is going to have her work cut out with you around, I can see.'

Days in Withered Hill: 357

When Zeke Ellison brings out their ploughman's lunches, Sophie can see from the corner of her eye that the land-lord's son doesn't know which of their tops to try to look down first as he leans over the table and places the plates in front of them. Catherine meets her eye and winks at her, then pulls down the front of her vest, showing her cleavage a little more.

'Do we get a discount for giving you a flash?' she says.

Zeke's cheeks burn like flames, and he scurries off back into the pub.

Sophie laughs and takes a long drink of her cider from the sweating pint glass. 'You're awful, Cat.'

'He's a sweet boy,' says Catherine, lighting up her third cigarette since they arrived twenty minutes ago. 'I might give him a treat next Faunalia.'

Sophie feels suddenly crestfallen. 'I won't be here then.'

'Nor Samhain,' notes Catherine, looking coolly at her. 'Maybe not even Lammas.'

'You think?' says Sophie, stricken. 'So soon?'

Catherine shrugs. Is there something in her eye, some flicker of what Sophie is feeling? Some doubt, or disquiet, at the thought of her leaving?

'Maybe I could come back…' she suggests.

'Nobody comes back to Withered Hill once they've left.'

'Lammas, though…' That is August the first. It is the middle of June already. Sophie doesn't know if that is a long time, or not enough time. She has to make a bower.

'I don't even know what they mean,' she says, glancing at Catherine. 'By a bower.'

'I can't tell you,' says Catherine, a little brusquely, then she seems to soften. She stubs out her cigarette and takes both Sophie's hands in hers. 'I can't tell you,' she says more gently. 'I can't. If I could, I would. I've helped you as much as I can while you've been here. But not this time. Not if you really want to leave.'

Sophie takes her hands from under Catherine's and places them on top. 'Maybe I don't want to leave.'

Catherine laughs. 'You've spent the last year trying to escape.' She takes a mouthful of cider. 'You held a knife to my throat. Have you forgotten that?'

Sophie looks away, at the cobbled main street. A delivery van has pulled up outside the butcher's. She watches as the driver and his boy unload a brace of pig carcasses from the refrigerated interior. 'No. No, I haven't forgotten.'

She feels Catherine's cool palm suddenly on her face, and looks into her green eyes. 'You have to leave. It's the way of things. You came to us so you could go. That's how it works.'

Home, thinks Sophie. She has a better understanding of that now than she did on that first day, standing on Carol Mountjoy's bathmat. But not for much longer. Soon, home will be outside. Just as soon as she finds out what she's supposed to do.

4

Outside

Days to Withered Hill: 12

Sophie was late. Again. She left the flat without even having a shower, which she knew was going to be a mistake as soon as she squeezed onto the grimy Tube as though she was being subsumed into London's morass of great unwashed, shorn of her protective aura of soap and shampoo. She'd dragged a brush through her tangled dark hair, and at least she'd cleaned her teeth for the requisite two minutes. Never again. No matter how drunk she got in future, she was not going to hit snooze so many times again and miss her shower before work. She resolved to set her alarm and put her phone on the other side of the room, but she knew she'd cave in almost immediately, when fitful sleep buoyed her into groggy half-wakefulness and she bathed in the pale glow of the screen to see who was saying what, who was liking whose posts and pictures.

She patted the reassuring weight of her phone in the pocket of her beige mac as she was ejected from the carriage, crushed up against shuffling bodies. Usually, she kept right on the escalators, lost in thought as she was inched upwards, but today she felt like she needed the sky, the sun, the breeze. She also needed food, her rumbling

23

stomach reminded her, and once she was on the street and could breathe again, she stopped off for a latte and crushed avocado on toast to go, mentally sticking two fingers up at, well, everyone. All those who shook their heads and said that if she only cut out the coffees and breakfasts she would be able to afford her own house.

Her gran kept asking when she was going to settle down. 'How old are you now, Sophie? Thirty-two? I had five children at your age.'

Yeah. And a job wherever and whenever you wanted one. And free university education, if you'd ever been bothered to leave the town where you were born. The glory years of the National Health Service. And a house that cost a thousand pounds. Sophie spent a grand before her wages had settled in her account properly on payday, most months.

She put her head down and weaved in and out of the flow of human traffic that all seemed to be moving in the opposite direction to her. Her phone buzzed in her pocket, and she transferred the paper bag with the half-eaten toast to the one holding her coffee, and reached for it. Sophie only had notifications set for messages from her one hundred and fifty-nine followers. It was probably her gran, who had recently discovered social media, typing in her weirdly Noughties teenage textspeak. *R U eating properly? Any chance 2 meet a boyfriend?*

Before she could tap the app open with her thumb, she walked into a brick wall.

Well, not a wall. A man.

Her coffee splashed all over her jacket and blouse and she staggered backwards, dropping her phone. The heel on her right shoe bent and snapped and she felt her ankle twist, couldn't keep her balance, and fell on her backside, the half of the coffee that was still in the paper cup dumping itself in her lap.

'Oh. My. God. I am so sorry.'

Sophie's fury was like a fever, making her shake and burn. She was covered in coffee. Her half-eaten breakfast lay on the pavement, crushed under the boot of an oblivious workman talking loudly on his phone.

'You fucking idiot!' She seethed, looking up at the man who had crouched down in front of her.

'I'm sorry. I wasn't looking where I was going.' He put his head on one side. 'Then again, neither were you.'

Her anger, which had started to dissipate as soon as she noticed that this guy was actually hot, rose again like the mercury in a thermometer. 'You're saying it was *my* fault?'

A man with shiny black shoes and a pinstripe suit tutted at her as he almost stood on her hand, flat on the pavement, and the hot guy – expensively cut hair, nicely trimmed beard, slight cragginess to him, piercing blue eyes – offered her his hand. 'If we're going to argue about this, maybe we should do it standing up.'

She accepted his hand grudgingly and let him pull her up to her feet. He bent down to retrieve her phone and straightened up. Sophie was five-eight and he was a couple of inches taller than her, even in her heels – one heel, she corrected herself, as she wobbled on the broken one. *Stop sizing him up*, she admonished herself. *He's not boyfriend material. He's some wanker who's just ruined your day.*

He frowned and bit his lip as he looked her up and down. 'Oh dear. I've made quite a bit of a mess of you.'

She looked down and groaned. The front of her white shirt was sodden, the spreading stain rendering the cotton see-through, clearly revealing her low-cut white bra beneath.

'Great,' Sophie muttered. She was now late for work, smelled like a tramp, and looked like shit. The perfect Monday morning.

'Look,' he said. 'I should pay for this. The dry cleaning.' He pulled out a brown leather wallet and riffled through it. 'Except… no cash. Can I ping you the money or something?'

'It's fine,' sighed Sophie. She never bought clothes that needed dry cleaning anyway. Fast fashion was Sophie's thing; buy it cheap, wear it out, and throw it away.

He scratched his beard. 'No, I should do something. I insist.' He dug into his wallet again. 'I won't ask you for your number because that would be weird. But take this. Call me or send me a message with the best way to pay you.'

She took the business card and shoved it in her pocket without looking at it. 'OK,' she said uncertainly. 'Thank you. I will.' She had no intention of doing that at all.

Then she hurried past him and into her building, casting a glance over her shoulder as she pushed against the revolving door. But he had been swallowed by the crowd.

–

Somehow, Sophie made it through to lunch after three hours of data processing. She had no idea what the data was that she was processing, and didn't care. Strings of numbers and letters. It obviously meant something to somebody. Was probably making people pots of money, on the back of her toil. Still, as Colin had promised her that night in the pub almost three weeks ago, it wasn't difficult. And it paid better than the minimum-wage, zero-hours temping jobs she'd been bouncing between for the past decade. Sometimes she amused herself by trying to spot words in and among the gibberish, awarding herself points for them as though on a game show.

She went into the bathroom, where she unbuttoned her coffee-stained blouse and rinsed it out under the tap, holding it in the hot air of the hand dryer until the stain was barely visible. She rolled deodorant under her arms, then liberally sprayed a cloud of Chanel No. 5 and walked in and out of it, like she'd read you were supposed to do. Luckily she'd been on enough nights out and lunchtime drinks to always have a spare pair of shoes under her desk. She still looked and felt like a crash between a bin wagon and a manure truck, but that didn't stop Colin from sliding over to her desk like a slug, offering to buy her a coffee and asking if she had plans for lunch.

When she'd taken the job at Gemini he'd immediately latched onto her. It was difficult to avoid him. Most days, there was only him and her working here, and Amanda – Mandy, she insisted Sophie call her at the interview – in the office. At least he'd had a half-squashed tube of glue in his desk drawer, though, so she could effect a temporary fix on her shoe heel.

'I've got plans, sorry,' she said curtly, signing out of her computer. A big clock appeared on the screen, counting down the thirty minutes she had for her break. If she wasn't back to log in again before it registered a bank of zeroes, it would go on her file as an extended break, and she'd lose an hour's pay.

They weren't allowed to use their phones at their desks, apart from in emergencies, so as soon as she left the building, she clicked on to her socials, remembering the notification. It was a direct message from someone called

@coldiron6239745, who she wasn't following, so she had no idea how they'd managed to DM her. It said simply, THEY ARE AFTER YOU.

A bot, or a nutter. Either way, she wasn't going to engage. The former and she'd probably end up with a ton of viruses on her phone. The latter, and the next message would be a dick pic. Still, it made her shiver a little bit, in the warm, early-summer sun.

THEY ARE AFTER YOU. What sort of sicko would send that to a complete stranger? She went to delete it.

Actually, she thought, *I'll report him.* Because she was fairly certain it was a him. But when she clicked on his name, it went to a page that said *Profile not found.* Sophie shrugged and swiped the message into the bin. Probably someone else had reported him first and they'd already suspended his account.

Sophie bought a sandwich and a latte at Pret, holding her breath, crossing her fingers and silently praying as she flashed her bank card at the terminal. It went through. She hadn't dared look at her balance for five days now, and there were still four days to payday. Provided she kept her job, of course. She was on a month's probation. Coming in late this morning looking like something a cat had dragged in through the cat flap probably wasn't going to do her any favours.

She remembered that the guy who had walked into her had given her his card, and as she hurried back to the office, her ankle still sore from where she'd gone over on it, she dug it out of her pocket. It was matte black and faintly embossed and had just a name and a phone number. No corporate logo, no job title, no email, no social media. Just *Tom Gisburn* and his mobile.

Tom Gisburn. Well, either he had no job and just liked to hand out his number to women on the street, or he had the sort of job that was so important he didn't need to put it on his card. On the basis of his slim-fitting suit, crisp white shirt and expensive cologne, she suspected the latter.

She took another look at the card, intending to crumple it up and throw it in the rubbish bin she was passing, then decided against it and pushed it back into her pocket. The last time she'd been given a business card something good had come of it.

Sophie got in with five minutes to spare, only to find Mandy standing at her desk, arms crossed. What had she done now? Sophie shook off her coat and said, 'I'm not late.'

Squat-nosed, hatchet-faced Mandy nodded to the desk, on which was a long black box, the sort that might contain a bottle of something. 'That came while you were out. You know the company rules. You're not allowed to accept gifts from clients.'

'I don't know anything about any gift,' replied Sophie, picking up the box. There was something heavy in it that slid to the end as she upended the box. Almost certainly a bottle. There was a card taped to the box that had her name and the address of the company on it.

'Open it, then,' urged Mandy. 'I'll need to log it in the system.'

Sophie shrugged and laid the box on its long side, picking at the tape that held the flip-top lid down. She opened it up and Mandy leaned forward to look. They both screamed and Sophie pushed her wheeled chair back, slamming into Mandy behind her.

'That's disgusting,' said Mandy, wrinkling her nose. Sophie could feel the supervisor's suspicious glare boring into the back of her head. 'Why have you been sent that?'

Sophie risked another glance inside the box at the dead, glassy-eyed rabbit.

'I have absolutely no idea,' she replied. She looked over her shoulder at Mandy. 'Do you still want to log it?'

–

After work, Sophie went straight to the Little Angels charity shop. They normally closed at five thirty, but tonight was a big stock-taking session. Sophie had said she would help out, which she was regretting now, after the day had left her feeling lethargic and unsettled. But they'd be expecting her.

The shop was run by Kath and Winnie, two women in their sixties, as different as chalk and cheese, as her gran would say. Winnie was loud and seemed to occupy all the space she stood in, always laughing and telling stories of her seemingly endless supply of Jamaican relatives. Kath was thin and bird-like, quiet, at least compared to Winnie. When Sophie let herself into the shop, they were arguing good-naturedly about politics.

'What do you think about him?' bellowed Winnie, mentioning the name of some politician Sophie only half knew.

'Oh, leave the girl alone!' said Kath. 'She's just done a day's work. It's good of her to give her time up like this for us. She doesn't want to be bothered with politics.'

Sophie smiled and went to put the kettle on in the little galley kitchen at the back of the shop. When she'd gone into Little Angels the first time, on the spur of the moment when passing, and volunteered her time, Winnie had asked her why, if her life had ever been touched by the loss of a child.

Sophie had said no.

5

Outside

Days to Withered Hill: 11

Sophie dreamed that she had a baby. She held it, swaddled in knitted blankets, in the living room of her flat. But now it was a nursery, with brightly painted walls and a rocking chair and a white crib. There were rainbows and unicorns on the walls, and the room was flooded with light from the windows that now looked out not onto a busy road, but fields and meadows and hills that met the blue sky at a distant horizon. When Sophie turned back from the window, her gran was sitting in the rocking chair, a beaming, satisfied smile on her face. 'I'm glad you've come home,' her gran said. 'It's good that you've settled down at last.'

'But this *is* my home,' said Sophie in her dream. 'My flat. I already live here.'

Gran smiled, and rocked, and watched Sophie lay the baby down in the cot. Except when she moved the blankets away from its face, it wasn't a baby at all. It was a rabbit. A dead rabbit. Not like a rabbit from one of Sophie's picture books, all wide-eyed and smiling. Its fur was matted and greasy, its body caving in, rotting and soft. Its eyes stared blankly, glassily, covered with a milky film.

Hare, she reminded herself as she sat up in bed, elbows resting on her knees, her hands cradling her face. Her duvet was drenched in sweat. It hadn't been a rabbit after all, it had been a hare. Colin had helpfully pointed that out. Sophie didn't much care; it was dead, it smelled and someone had sent it to her at work, which was the most disturbing thing.

Mandy had wanted to call the police, but Sophie had said no. Even if they took it seriously, what would happen? They'd quiz her for hours to try to find out who might have sent it, and that would involve dragging up everything with Jamie again, because, on paper, he was the likeliest candidate. But even Jamie wouldn't do something so gruesome, Sophie was sure. She hadn't heard from him for eight months now, and the last thing she wanted was the police putting her back on his radar, when, in all likelihood, he'd shifted his obsessive attention on to some other poor girl. Besides, if it wasn't Jamie, that meant that there was someone else out there who had some weird fixation with her, and she really didn't want to think too deeply about that. At the end of the day, it was just a dead rabbit.

Hare, she corrected herself. It was just a dead hare.

Sophie had woken up in time for a shower this morning, thanks to the dream shocking her out of sleep early, and she stubbed her toe on an empty Pinot Grigio bottle as she swung her legs out of bed, her head hammering. That bottle had been almost full when she went to bed. The glass was on its side by the bottle, a still-damp patch of wine spreading out on the carpet. At least she hadn't finished the whole bottle, she thought. She'd left at least half a glass full. Of course, there was the other empty bottle sitting judgmentally on the draining board when she went to get a couple of ibuprofen and a glass of water. Still, after the day she'd had, who could blame her?

–

Sophie arrived at work feeling much more human than she had yesterday. She went to her desk and started on the day's data entry workload. Her in-tray held a sheaf of paper with handwritten strings of numbers, all zeroes and ones. She took the top sheet and her fingers danced over the keyboard, her eyes skimming the lines of numbers, her brain sending them as signals to her hands, which unpacked and replicated them on the computer monitor. It was only when her monitor buzzed angrily and the screen froze, to be replaced by the big thirty-minute countdown, that she realised she had been working for three hours straight. She hadn't taken a single drink of the takeaway latte that sat cold on her desk, nor spoken to anyone, nor gone to the toilet. She had been lost in the numbers, almost in a trance. It was as though her mind was fogged so much that she could barely remember her name or what she was doing there. She had read something about that before. As the murk cleared, it came to her. *Fugue state*. Total disassociation with one's self. A feeling of travelling and being lost. Not knowing who you were.

She had already wasted two minutes of her lunch break and she stood quickly, gathering her bag. She was incredibly hungry all of a sudden, which wasn't really surprising as breakfast had been two painkillers and a pint of tap water. She turned to leave, but Colin was blocking her way.

'I did a bit of research for you,' he said. 'On hares.'

She frowned at him. 'Why?'

'In case you wanted to know why it had been sent to you,' he said, a little uncertainly. 'Don't you want to know?'

'Not really. I just want some lunch.' She paused, suddenly curious. 'What sort of research? What did you find out?'

Colin shrugged. 'They're associated with the moon in lots of cultures. Chinese, Greek, Celtic. In English folklore, they represent sexuality, fecundity and rebirth.'

'Fecundity?' said Sophie.

'You know, like fertility,' muttered Colin. 'Hares appear a lot in folklore. Witches' familiars, sometimes witches themselves, shapeshifting.'

Sophie thought about her dream, the dead hare swaddled in baby clothes that she thought was her child. Witches, fertility, rebirth. She felt a little shiver in the air-conditioned office.

'Well, thank you, Colin. That's all very helpful.'

'Really?' said Colin, a hopeful smile splitting his doughy face.

'No, not really,' replied Sophie. 'If you'll excuse me…'

She bought herself a sandwich and a coffee with an extra shot of espresso, then went to Boots to get more ibuprofen. Her headache was coming back, and things hadn't been helped by entering all that number data in the morning. It was another sunny day in London, but she could almost feel the high pressure behind her eyes. The shop signs seemed to swim in her vision, the letters shifting into zeroes and ones and back again. She spent too much in Boots, had spent too much everywhere, it seemed, because the contactless payment wouldn't go through and she had to stick her card in the machine, staring at the keypad for ages as her PIN eluded her, her head full of binary code.

Sophie ate her sandwich on a bench, watching a pigeon stalking up and down in front of her, glaring at her with beady yellow eyes. She balled up the cardboard wrapper and threw it at the bird, and a thin woman in a twinset tutted at her. Sophie downed two ibuprofen with the cold dregs of her coffee and slunk back into the office.

She sat down at her desk, lifting the next file from the in-tray, immediately starting to type in lines and lines of what appeared to be random words in French.

It was after nine when Sophie got home, her fingertips numb from typing. She slung her bag on the sofa and went to the fridge, dropping her coat on the floor and leaving it there, and stared at the meagre contents. She decided she couldn't be bothered to eat anyway and grabbed one of the two full bottles of chilled wine, her anxiety dissipating noticeably with the satisfying twist and crack of the top on the bottle, and the welcome glug of liquid into a large glass.

She flopped onto the sofa and switched on the TV, clicking on to Netflix. A message appeared on the screen. *Your account is on hold because of a problem with your last payment.* Sophie sighed and flicked it back to terrestrial. There was nothing on, so she settled for an old sitcom from the eighties, the ratio of the picture distorted to fit the widescreen TV. That was how she felt. Stretched and distorted.

The show had been made before she was born, and she found it unfunny and unfathomable. But she watched it anyway, or at least stared at it, until the glass of wine was finished. Then she went to the fridge and filled it up again, and sat down in front of what might have been the same episode of the same sitcom, or something completely different, and just let it all wash over her.

—

She woke at just before three in the morning, her neck aching and stiff from where she'd dozed off on the sofa. The TV was showing some kind of all-night gambling programme, a red and black roulette wheel spinning hypnotically, the names of people who were haunting the small hours in the hope of winning big running across the bottom of the screen. Perhaps she should have a bet. Put a tenner on. You never knew. But the roulette wheel was blurred and indistinct, all the numbers seemingly zeroes or ones.

She reached for her phone and saw two notifications.

One was a direct message from @coldiron6239745, evidently out of social media jail for whatever infraction he had committed. It just said THEY STILL WANT YOU. YOU SHOULD RUN. Sophie clicked on the profile, but it was dead again. She blocked him anyway and deleted the message.

The other notification was for a text, from a mobile number she didn't recognise. It said *How about Thursday? I'll buy you a drink and a bite to eat if you'd do me the honour.*

She puzzled over it for a moment until she realised it was a reply to a text she'd sent, sometime just after midnight. She had no memory of it, and her hand flew to her mouth as she read it. *Hey, Mister Clumsy, you going to pay for my ruined clothes or what? Want to meet up? I'm free all week.*

Sophie stood unsteadily and retrieved her coat from the kitchen. The business card was gone from her pocket. After some casting around, she located it on the kitchen worktop, next to an empty bottle. At first, the phone number looked like ones and noughts, but she blinked that away and compared it to the message on her phone. It matched.

She groaned. She had drunk-texted him. Tom Gisburn. And he wanted to meet her.

Sophie went to the toilet, threw up noisily, and climbed into bed for four more fitful hours of sleep haunted by dreams of gigantic zeroes and ones that stomped across the English countryside, crushing squealing hares beneath them.

6

Inside

Days in Withered Hill: 46

After almost seven weeks, Sophie has given up – at least temporarily – trying to escape. She has been confounded at every turn, turned back at every attempt. The long days bleed into each other, typified by sunshine and blue skies and a growing sense of lethargy, and not just because of the summer heat. Sophie does not quite consider Withered Hill to be home, as she was repeatedly told it was when she arrived, but in the absence of a single memory of any other place she can rightly call home, it will have to do.

Everyone is so… nice. Sophie doesn't understand. She is a prisoner here, in that she cannot leave. But no one threatens her. No one seems to wish her harm, or even gives her a cross word. And nothing seems to be expected of her.

Until Lammas.

Sophie is sitting outside her cottage, sunning herself and reading a book lent to her by Mr Obermann at the library on the Pendle Witches, when Catherine pulls up in her little Mini. Sophie pulls down her sunglasses to watch Catherine climb out of the car, tanned almost mahogany by the endless sunshine. She pulls the seat forward and leans into the back, before turning round, ducking her head out of the small car, and saying, 'Are you going to sit there staring at my arse or are you going to help me?'

Cheeks burning, because she has still not got the measure of Catherine, can never tell whether she's joking or is genuinely annoyed, Sophie rushes over and mumbles an apology. 'I wasn't looking at your... at you.'

Catherine raises one eyebrow, her eyes dancing with the smile her mouth keeps hidden. 'I'm winding you up. Give me a hand with this.'

Grunting, Catherine manoeuvres a large cardboard box out of the car and Sophie takes one end to help her pull it out.

'Have you been outside?'

'Not today,' says Catherine. 'This is from the school.'

It isn't heavy, just bulky. They place the box on the dry grass and Sophie pulls at the loose flaps. Inside are masks painted by the children, presumably before the summer holidays started. Some are rough, some beautiful. They are all hares, big enough to cover an adult face, eye holes carefully cut, a length of elastic stapled to hold the mask secure.

'What are these for?'

'The festival tomorrow night.'

Sophie picks one of the masks up, rubbing her finger over the waxy crayon lines. 'Festival? Wasn't that a few weeks ago?'

Catherine laughs. 'That was Midsummer. Tomorrow is Lammas. You'll get used to it. We have a lot of festivals in Withered Hill.'

Sophie has only been in Withered Hill for a week when Midsummer comes round. Stung by her repeated failed escape attempts, she stubbornly refuses to attend, and locks herself in the cottage they had just moved her into. One by one, they come to the door, knocking and calling through the letter box.

'Don't be a silly goose!' trills Carol. 'It's fun! Like a big party!'

Sophie tells her she isn't in the mood for a party.

'Get yourself out of there at once, young lady,' orders Zachary Winterbottom.

Sophie ignores him.

'Don't come if you're going to be like that,' says Catherine through the door. 'You'll only be a downer for everyone else.'

Sophie's stubbornness almost pivots 180 degrees, and her interest is piqued, but she just tells Catherine to enjoy her Midsummer party and turns back to her book, a tattered old paperback romance set during the war.

All evening, Sophie sits in the chair, reading first by the sinking sun, then by lamplight. It is stiflingly hot in the cottage, and she opens the windows, listening to the sounds of music and revelry floating on the warm, thick summer air. Even when it gets properly dark, the last day of the year that the moon would rise so late, the sounds of the festival continue. Sophie wonders what is going on there but continues to read until her eyes start to droop and her head nods.

She awakes sharply, not sure how long she's dozed. She blinks, and yawns, and lays down her book, then starts. There is a figure outside, a little way in front of her cottage, standing stock-still, as though frozen by moonlight.

It is Catherine. She is just stood there, staring at Sophie's window, a wine bottle held loosely by her side. Sophie stares back at her, for what feels like an age. Catherine doesn't move. If she wants to speak to Sophie, why doesn't she just come to the cottage?

Frowning, Sophie stands and walks to the front door, wrenching it open. The spot where Catherine had stood is empty, and Sophie can see no sign of her in either direction.

–

'I'm assuming you're not going to sulk through Lammas like you did through Midsummer,' says Catherine, hefting up the box and walking to Sophie's cottage.

'I assume you're not going to stand outside my window again just staring at me, if I come,' replies Sophie, following Catherine inside.

Catherine grunts, and says nothing else, putting the box on the table.

Sophie asks, 'Tea?'

While she's making a pot, she calls from the kitchen, 'So what happens at Lammas, then? What does that even mean?'

'It meant loaf-mas, originally,' says Catherine, adopting her teacher voice. 'It was a Christian festival, though far older than that really. On the first day of August, we give thanks for the grain and the bread, and for the harvest to come.' She watches Sophie place the tray of tea on the small table in front of the sofa. 'It's my least-favourite festival.'

'Why?' asks Sophie, pouring the tea.

'The bloody carbs,' replies Catherine, patting her stomach.

Sophie laughs. 'And the masks? Rabbits?'

'Hares. Symbol of fertility and all that. Very special.' Catherine puts her head on one side. 'Though the farmers hate them. Ask Peter O'Keeffe. Pests, they call them. There's been a lot of shooting, trapping and coursing this week.'

Sophie picks up a mask and pokes her finger through the eyehole. It's designed to sit on the face but leave the mouth and nose exposed. 'Revered, yet killed as pests. That doesn't make sense.'

Catherine shrugs. 'Those are the choices that have to be made when you're custodians of the land like we are in Withered Hill. It's about balance. It's entirely possible to be one thing and yet at the same time be another completely. Got any biscuits?'

Sophie goes to get a packet. 'And why have you brought the masks here?'

'It's your job to distribute them around the village. Tonight. Ahead of the festival tomorrow.'

'My job? Why?'

'You're the Child of Promise,' says Catherine with a shrug.

'Am I? And what does that mean, exactly?'

'I'm sure someone will tell you eventually.'

–

As instructed, when dusk falls, Sophie heads into the main street of the village with the box of masks. There are three men from the pub at the bottom of the cobbled rise, one with a fiddle and one with an accordion. The third begins to sing in a deep, baritone rumble, a jolly, bouncing song that Sophie can't make out the words to. The fiddle player nods at her, and she begins to walk up the street.

As she passes each door, they open, revealing one of the villagers holding a flickering candle. Wordlessly, Sophie passes a mask to each person, who accepts it smiling but silently. When she gets to the top of the hill, near the Post Office, Carol beams at her from her doorway, holding a stout altar candle, and Sophie hands her the last mask in the box.

'Well done, my lovely!' whispers Carol, and closes the door.

Far behind Sophie, the music tails off and stops. She is standing alone in an empty street, darkness descending. She waits for a while, but no one comes, so she walks back to her cottage, wondering what the next day will bring.

–

Catherine calls for Sophie at a little after seven in the evening, and they walk together to the town square. There is a delicious aroma of baking bread in the air, which makes Sophie's stomach rumble.

'Has everyone been making bread?'

'Most people,' says Catherine, nodding. 'The wheat is the first of the harvest. It's been a good year, as usual. Lammas is our way of saying thank you.'

'Thank you to who?' wonders Sophie aloud.

Catherine just smiles.

Sophie says, 'I suppose that's something else I'll find out when someone decides I'm ready to be told.'

'You got it. That's the—'

'I know. That's the way of things in Withered Hill.'

'You're learning.'

'It would be difficult not to. I think everyone in the village has said that to me at least three times since I arrived.'

Since I arrived. Over the last seven weeks, Sophie has had much time to reflect on her arrival in Withered Hill. An arrival suggests a journey, and a journey indicates a starting point. She cannot remember where she came from, or why. That said, with each passing day, vague memories surface, sometimes gone as quickly as a puff of smoke, others that settle and last. But they are like pieces of a jigsaw and she has no idea what the picture will reveal. Not yet.

At the village square, there are long trestle tables set out around the edges, each one groaning with steaming loaves, fresh from the ovens of Withered Hill. There are tables of beer and wine, and music from the three men who played the previous night as Sophie handed out the masks. Which are all now on Withered Hill heads, the square a sea of crayoned and poster-painted hare faces, bobbing along in time to the music, the masks getting soggy from the drinking.

Catherine reaches into her bag and pulls out two masks, one for her and one for Sophie. Sophie follows her lead and pulls the elastic over her head, peering through the roughly scissored eye holes.

Behind the mask, she feels less exposed, more a part of Withered Hill and what is going on, though she has no understanding of what that might be, yet. The conversation suddenly dies and the music fades. There is a squealing of feedback and Sophie turns, as does everyone else, to face in their hare masks a small platform on which stands Noah Jones, the chairman of the parish council. He is holding a microphone, and there is laughter as it taps loudly against his head, amplified by the speakers, as he pulls on his own mask.

'It is Lammas, and we give thanks,' says Noah gruffly into the microphone. 'Thanks for the harvest we hope to receive, thanks for the land that looks after us, and those that are the earth and are of the earth. Aye, you all know who I mean.'

There's a rumbling of assent from the villagers.

Sophie glances at Catherine, but her eyes are on Noah, who says, 'Now it is time for our Lammas celebrations proper, so take thee the fruit of the fields and let us sing our joy and gratitude that we live in this paradise we call Withered Hill.'

A cheer erupts from the crowd and they turn and move towards the spread, descending on it, grabbing at the loaves piled high.

Catherine grabs Sophie's hand and drags her towards the nearest table. People are already moving away from them, biting down on the warm bread.

'Here,' says Catherine, pushing a loaf into Sophie's hands. They are being jostled by villagers hungry for the bread.

'What? We just… eat it?'

But Catherine is already biting into the loaf, her cheeks becoming swollen as though she has starved herself for a week beforehand. Her eyes are shining with something Sophie hasn't seen in her before. She nibbles at her own loaf, glancing around at the villagers in their paper masks.

All she can hear is the sound of chewing and biting and the occasional groan of pleasure. Sophie looks at her loaf. It's huge. She can't be expected to eat all of it, surely? She glances up at Catherine, who's already gnawing deep into the centre of the bread. Then she lifts her head and smiles broadly, and Sophie's eyes widen.

Catherine's mouth and chin are slick with something dark. Sophie looks around at the other villagers. They're all the same. There's something in the bread. Something that's… not bread. Frowning, Sophie begins to sink her thumbs into the loaf, to prise it open. Catherine is staring at Sophie, eyes wide and wild. There's an audible crunching sound from her mouth and she spits out a mouthful of what looks like chewed bread and sticks. Wait, no, not sticks. Thin white bones. Her face is getting wetter and stickier and…

…And bloodier. It's blood. In the bread. There's something in the bread and…

Sophie screams and drops the loaf. Everyone ignores her and continues to eat, as though each mouthful is making them more ravenous, not sating their hunger as it should. Sophie stares down at the torn-apart loaf, and what sits inside it.

The skinned, baked, but unmistakable head of a hare.

Feeling her bile rising, Sophie pushes away from Catherine, elbowing through the crowd, and runs for her cottage.

7

Inside

Days in Withered Hill: 358

'Sophie!' says Mr Obermann warmly as she walks into the cool shade of Withered Hill Public Library. He is sitting at the desk, stamping books with a satisfying clunk of the ancient device. He presses first onto a pad of red ink before bringing it down on the library insert on the flyleaf. Behind him, dimly lit aisles of shelves disappear into shadowy vanishing points.

Sophie likes the smell of the library; it is dusty and old and simultaneously full of new and fresh things to learn, as though the shimmering motes that dance in the stained-glass sunlight are not just dust, but stardust.

Once, long ago, the library used to be a church, but they have no use for churches in Withered Hill, haven't done for a long, long time. In fact, the building was never really finished before the church gave up on Withered Hill. It was meant to have a spire, Mr Obermann told her once, but the place had been abandoned long before it was ever completed. There was the start of one, jutting up from the rear of the building, but it had been capped off and at some point someone had put railings around it. Now it was a flat space that Mr Obermann had turned into a roof garden with big potted palms and magnolias and overflowing troughs of tomatoes and carrots. Things grow well in Withered Hill, given that it is in the heart of the Lancashire countryside. Crops are bountiful, livestock is fat, villagers are healthy and strong. Fecund, thinks Sophie. That is a good word for it.

When she had first arrived, she had asked why the village was called Withered Hill, when it was obviously anything but. The woods that circle it are thick and dense, and the farms harvest groaning carts of gleaming, swollen produce. Mr Obermann told her that it was a derogation; a name coined in the mists of long ago by the other villages around Withered Hill, jealous of its bounty.

Sophie is wearing a pair of denim shorts and a black vest that had arrived in her package yesterday. Mr Obermann, with his close-cropped white hair and lined, weathered face, gives her an appreciative glance from her black hair to her pump-shod feet. 'I'll be sorry to see you go, Sophie. You always put lead in this old librarian's pencil.'

She blushes a little, which Mr Obermann catches, and he laughs.

'Pay me no mind, girl. Just the wittering of a lusty old man. The day I wake up and don't have the horn for a pretty, young girl, that's when you can cart me off to the barrows.'

It is the way of things, in Withered Hill at least. Sex is not something to be ashamed of, or hidden away, or spoken of furtively.

Mr Obermann seems to be reading her mind, and says, 'They are the first numbers, the zero and the one, and it is only natural.' He makes a circle with the forefinger and thumb of his left hand and pokes his right forefinger into it. 'The zero and the one, you see? It's just mathematics, really.' Mr Obermann stamps the final book on his pile and closes it. 'Though I am given to understand that things are very different on the outside these days.'

'Mr Obermann...' begins Sophie.

He stands up, his bones creaking, and he rearranges the crotch of his trousers to accommodate the bulge there. He winks at Sophie and says, 'Not ready for the barrows yet, this old goat. But I imagine you come for my purer thoughts.' He taps his temple with a finger. 'I have to tell you, I cannot help you directly. But I can guide you a little.'

'I guessed that,' says Sophie. 'I just want a starting point, really. With this bower business.'

Mr Obermann reaches to the end of the desk for a fat, hardback encyclopaedia. 'Let us see, then,' he says, licking a finger and flipping over the pages. 'Ah, here we are. Bower.' He glances at Sophie. 'Three definitions. One, a pleasant shady place under trees or climbing plants in a garden or wood. Two, a summer house or country cottage. Three, a woman's private room or bedroom.' He slams the big book shut. 'Bower. There we have it. Does that help?'

She wrinkles her nose. 'Not really.'

'Come with me,' says Mr Obermann, and he leads her to an arched, mahogany door at the back of the library. Sophie walks behind him, her fingers trailing the spines of the books, as though she might absorb their knowledge by osmosis to help with her conundrum. He takes a bunch of keys from his belt loop and unlocks the door, opening it to a narrow stone spiral staircase. 'Ladies first,' he insists.

Sophie climbs the staircase, holding on to a rope bannister set into rings in the cold stone wall. At the top, there is a narrow landing and a hatch set into the low roof. Mr Obermann squeezes past Sophie, reaches up to pull at the wooden hatch, and a ladder slides down to their feet. Sophie says, 'Ladies first, I suppose?' then climbs the ladder, emerging out onto the flat stone roof of the almost-spire.

Far to the east, white clouds with bruised, grey undersides scud along the horizon, but otherwise, the blue sky is unblemished.

Mr Obermann follows her gaze. 'They'll be from Yorkshire. Dour lot they are over there. Did you know that it was once said Lancashire is the country where women die of love?'

Mr Obermann's roof garden is the best vantage point in all of Withered Hill. You can see the entire village from here: the cobbled main street below her, the warren of side streets, the cottages and bigger houses spreading out, then the fields and farms and finally the thick ring of woodland that surrounds the village. Beyond, there are two roads, one heading east towards the blackening clouds, the other west through abundant moorland. She frequently sees the farmers' vans heading out in either direction, taking their produce to the markets in Skipton and Clitheroe and sometimes beyond. Withered Hill beef especially is a delicacy much sought-after in the Michelin-starred restaurants of the Three Counties, famed for its tenderness and plumpness. In the butcher's shop, there is a yellowing cutting taken from a national newspaper, in which a famous restaurant critic waxes poetic about a Withered Hill-grown steak he ate in the Trough of Bowland, likening it to an almost transcendental experience with hallucinogenic qualities. Mr Purcell, the butcher, once watched her reading the cutting and said to her that the critic had gone mad afterwards and was in an asylum somewhere. 'Once you've had a bit of Withered Hill in you, you never get rid of it,' he'd said, with a wink.

'Beautiful, isn't it?' says Mr Obermann, joining her beside a tall, bushy olive tree. They are looking west, hills and moors as far as they can see. Beyond it all, though, Sophie knows are other villages, towns, cities. And the sea.

'Which way's London?' she asks.

Mr Obermann puts his hands on her hips and steers her to the left, and points south. Beyond the trees lie the barrows, where Withered Hill's dead are taken to their final rest. 'Big place, London. Noisy. Busy. Dirty. Unfriendly. Nothing like here. But I don't need to tell you that.'

'I'm not sure I want to leave here and go there,' murmurs Sophie.

Mr Obermann laughs, but not unkindly. 'You've changed your tune. I remember a time when you would have done anything to get out of Withered Hill.'

Days in Withered Hill: 1

Carol Mountjoy finds her a floral summer dress, short at the hem and low-cut at the top. 'Show off that lovely figure,' she says, unfastening a button at the cleavage.

'Why does that matter?' says Sophie.

'Because that sort of thing is important out there. They put more stock in it than kindness and brains, even if they say they don't.'

Sophie frowns. 'Out there?'

Carol stands back to appraise Sophie, who she has given a pair of white pumps to wear. 'I think you'll do. Let's go meet the others.'

They walk downstairs and out of the Post Office, onto the cobbles of Main Street. There are more people out now, and they all look at her with unabashed interest. None of them are wearing animal heads or masks though, and she wonders if she dreamed all that. But then she sees the butcher talking to a woman in the street, and his white tunic is stained around the shoulders and collar with blood.

They pass more shops and small cottages, Carol talking away merrily, though Sophie isn't listening. Then they come to a large, stone-flagged square with a maypole at the centre, wound about with coloured ribbons that are tied together at the bottom and flap against the pole in the warm breeze. At the far side of the square is a grand building with columns on the porticoed frontage, and a hand-painted sign that says *Withered Hill Parish Council*.

Inside, there is a staircase that leads up from the marble-floored foyer to a landing covered with thick green carpet, off which there are five or six rooms behind thick oak doors. Carol opens one and leads Sophie into a room dominated by a big, circular table. On the wood-panelled walls are paintings, portraits of men in old-fashioned suits and landscapes of the countryside. There are some with scenes she cannot understand, devilish figures and naked women entwined in woodland clearings.

Later, she will come to know the five men and one woman seated at one end of the table, facing her. They are Noah Jones, the chairman of the parish council, Thaddeus Obermann, the librarian, Zachary Winterbottom, Michael Ellison, the landlord of The Farmer and Devil, Peter O'Keeffe, who runs Nut Nan Farm, and Emma Mangnall, who is the secretary at St Michael's School.

Carol pats her shoulder and tells her to stay standing, then goes and sits beside Emma.

Noah Jones looks along at them all, then raps the table with a small hammer. 'Withered Hill Parish Council is now in session.'

'Well, I must say they sent us a pretty one this time,' remarks Mr Obermann, smiling warmly at Sophie.

'What does she know of herself?' asks Noah Jones.

'She knows her name is Sophie,' replies Carol.

'Sophie Wickham,' says Sophie, suddenly plucking the name from the recesses of her mind.

Noah Jones makes a careful note in a ledger open in front of him on the table. 'Welcome to Withered Hill, Sophie Wickham,' he says.

She looks around, her eyes drawn to the paintings with the devils and women again. The sun shines through the big windows that look out onto the market square, and she can hear a child laughing. A dog barks. She feels a little bit unusual, as though the world is dimming at the edges. She says, 'Why am I here?'

'It's just the way of things,' replies Peter O'Keeffe, broad of chest, his big forearms black with writhing tattoos. 'You come to Withered Hill and we look after you.'

'Why am I here? Why can't I remember anything?' asks Sophie. She could do with a glass of water, or a sit-down.

'Why, all that will become clear,' says Emma Mangnall, lifting her glasses from where they hang on a chain around her neck to peer through the lenses at Sophie. 'My, she is a bonny one.'

'I told her,' says Carol. 'The boys'll go a bomb on her, I said. Catherine Baldwin's going to have her work cut out.'

'Why do you keep talking about what I look like?' demands Sophie. Her own voice sounds distant, as though heard across a wide valley. She taps her head. 'Isn't it in here what matters?'

'I told her, it's important on the outside,' says Carol.

'She's not wrong, though,' says Zachary Winterbottom. 'Maybe she'll not be with us all that long if she's worked that out already.'

As long as it takes, thinks Sophie. She knows this is wrong. She does not belong here. She walked into this village, this Withered Hill, naked and dirty and with no idea who she is or what she's doing here. And they think it's the most normal thing in the world, and that she is to stay here. As long as it takes.

Sophie turns towards the door and runs.

8

Outside

Days to Withered Hill: 10

Sophie ran. She had got home to a pile of parcels in the hallway stuffed with things she didn't remember buying. No wonder she was skint. ASOS, PrettyLittleThing, Boohoo, even a little weird corn dolly she'd apparently bought on eBay for a fiver.

She walked to the fridge and found it bereft of both wine and food, and glanced guiltily over to the empty bottles lined up by the kitchen bin like nine-pins. Among her late-night drunken purchases were a sports vest and a pair of yoga pants, evidently part of some misguided pledge towards self-improvement. She propped the dolly up on the mantelpiece and got changed into the gym gear, and dug out her Nike trainers from the bottom of the cupboard. Instead of going to the shop for wine, she would make good on her promise to herself, lost in a fog of alcohol that it was.

So, Sophie ran. She limbered up in the hallway of her block, stretching with one leg pulled behind her, then the other, and set off at a slow jog from her building towards the park. She got a hundred yards before she had to stop, leaning on a lamp post and trying to fight down nausea, stars pinging at the periphery of her vision. While she got her breath back, she downloaded an app on her phone, which, to her almost tearful gratitude, told her she didn't have to immediately jump into a half-marathon. Following the instructions in her earphones, she walked a hundred yards and then jogged a hundred yards, repeating the process all the way to the park.

These things come in threes, her gran used to say about bad luck. The dead hare and the weird messages certainly counted for bad luck. So she'd got up feeling anxious and sick that morning, waiting for the inevitable, and wondering what it was going to be. But work had passed without incident, refreshingly, boringly mundane after the last two days, and even Colin had kept out of her way. She had got her payslip and even though it didn't cover half of her outgoings, she would at least have money going into her account tomorrow and could barter and negotiate with everyone she owed so she could get through the next few days feeling like a normal human being.

After getting her breath back at the park gates, she began to jog around the tarmac path, nodding at other runners and sidestepping small, yapping dogs on leads. Sweat was beading on her forehead and the stitch in her left side was subsiding. She was actually starting to feel good as endorphins flooded her system. She resolved to run every single day after work.

Sophie did a full circuit of the park and decided to walk home from the gates. She didn't want to overdo it. Plus, she wanted to call into the shop, where she bought two bottles of wine and picked up a microwave ready meal, then put it back in favour of some fresh pasta, a low-fat sauce and a packet of pre-prepared salad.

Back at the flat, she stood under the shower, feeling the tightness of her leg muscles. It felt good. She dried off and then went into her bedroom, inspecting her purchases laid out on her bed. There was a lovely little, short summer dress with a bold tropical print. She tried it on and inspected herself in the mirror, then dug into her drawer for a matching set of pale mint underwear. They'd go nicely together, along with her cork wedges. She didn't try the underwear on, though. She thought she'd save them. For tomorrow.

Carefully taking off the dress and hanging it on the door of her wardrobe, she grabbed her phone and pulled up her texts. She'd already added Tom Gisburn's number to her contacts. After a moment's lip-biting hesitation, she replied to his message. *Thursday would be lovely. I'll be free from seven. Tell me where you'd like to meet.*

It had been forever since she'd been on a date. And nobody was fooling themselves here that this was about a dry-cleaning bill. Sophie lay back on the bed and picked up his business card, running her thumb over the embossed letters of his name. Tom. It was a good, strong name. Tom Gisburn.

She closed her eyes and recalled his icy blue eyes and his neatly trimmed beard, his cropped hair with the faintest suggestion of incoming grey, the strength of his hands as he pulled her up off the floor. She stroked the edge of his business card along the inside of her thigh, shivering as it raised goosebumps, imagining it was the tips of his well-manicured fingers.

–

After eating the pasta and sitting in her robe on the sofa with a glass of wine – she'd promised herself she'd limit it to one bottle tonight – she took the plunge and logged on to her online banking, peering at the balance through one eye. It wasn't quite as bad as she feared, and she had enough to pay the Netflix subscription and indulgently allow herself to browse Victoria's Secret, where she spotted a set of underwear that would be perfect with the dress, totally more so than the set she had. With free next-day delivery. Plus it was payday tomorrow and—

The phone buzzed in her hand just as she was checking out, and a notification slid down the screen. Text message from Tom Gisburn. She was more nervous about opening it than she had been about checking her bank balance. Eventually, she took a deep breath and swiped open her texts with her thumb.

> That's great! Do you know the Green Man on Langley Street? How about I meet you outside there at eight?

She did know the Green Man. In fact, it was only two streets away. She'd never been in because it was what she thought of as a drinker's pub. But it was close enough that she could get home without worrying about a cab if the date was a disaster. She replied immediately to say she'd see him there, then bought the underwear before time ran out for next-day delivery. Then she went to get a top-up of wine. The first bottle was empty apart from a tiny splash, so she opened the second. This was cause for celebration, after all. And after all that exercise, she deserved this.

–

When Sophie was twelve, she saw a ghost. That was the only way she could describe it. It had been at a church fete in the Gloucestershire village where she grew up. It was a scorching Saturday afternoon and on the field beside the church there were stalls and games and a beer tent, and a game of cricket being played, in which her father was participating. Her mother was helping out on a stall selling cakes and buns, and Sophie was bored, so she wandered off, beyond the church and into the graveyard behind it.

The graveyard was old and tumbledown and most of the headstones were weathered to the point of the carved inscriptions being illegible. It wasn't used as a graveyard any more; people were buried in the municipal cemetery in the next village. Sophie liked to walk the overgrown paths between the graves, sometimes spotting a squirrel bounding in the branches of the trees overhead, or even a fox nosing through the undergrowth.

But this time, she saw a ghost.

It looked like a man, except all hunched up and crouched down, perched on top of one of the flat grave-stones, ragged clothes flapping, like a blackbird. Its face was gnarled and white, as though carved from chalk, or the inside of a tree. She felt its eyes on her for a long time before she noticed it, when she had wandered just twenty feet away from where it sat.

'Tha's a pretty little 'un,' said the ghost. 'What's tha' name?'

Only a month ago, her mother had sat her down and talked to her about strangers, specifically strange men. Sophie had just started her period, and her mother acted as though she now had some kind of beacon flashing above her head, attracting the unwanted attentions of men. Sophie had noticed her body changing, her figure becoming curvier, hair sprouting between her legs. She had been given to understand that the sort of men who would be showing an interest in her were either faceless, eerie figures in coats and hats who would try to entice her into their homes, or tracksuit-wearing youths in fast cars who were 'only after one thing'. Sophie wondered what that one thing was. She wished people would be more specific about such subjects. The only thing she was pretty sure of was that gnarly-faced ghosts had never been mentioned to her.

That didn't mean she was stupid, though. She knew not to give a ghost, or sprite, or whatever it was crouching there on the grave, her name. She had read enough stories about that sort of thing. You didn't give something like that your name, or tell it where you lived, or accept an invitation of any kind from it.

She blinked and the ghost was suddenly right in front of her, though she hadn't seen it move. She gasped a little, and it grinned.

'Aye. Very nice.' It seemed to sniff the air with its big nose. It put her in mind of the Mr Punch puppet she'd seen at Weston-super-Mare beach that time. 'Reet young woman, aren't thee, now?'

Sophie took a step back. The sun had dipped behind the tall trees in the graveyard and it felt suddenly cold. 'I should be getting back to my parents.'

'Not before I have a kiss, surely?' said the ghost.

Sophie was pretty sure you should never kiss a ghost.

'No, sorry, I think I should go,' she said politely.

She turned round and started to walk back along the overgrown path to the church, suppressing the urge to run. She screamed as she felt clawed fingers rake her hair, and spun round.

The ghost was crouched on another headstone, laughing. In one hand, it held a pair of scissors, which it snip-snip-snipped. In the other, between its bent fingers, a hank of her dark hair dangled. 'This'll do in lieu of a kiss,' it said, and disappeared.

Sophie screamed and ran all the way back to the fete, and found her mother and told her what had happened. The police were called and the graveyard searched, but there was no sign of anyone. Sophie had to sit in a small room in the church with a lady police officer who gave her a doll with no face and asked her if the man had touched her, and to point to the places on the doll where he might have put his hands, and that she hadn't done anything wrong so wasn't to worry. Sophie pointed to the back of the doll's head and said, 'He touched me here. He cut my hair.' Her mother was called in and she inspected Sophie's hair, and she saw a glance pass between them, and her mother shrugged and shook her head silently. There was a CCTV camera on the church, installed after a spate of vandalism in the graveyard a couple of summers back, but the footage showed nothing other than Sophie standing by the graves, seemingly talking to herself, and then running away. Her mother told her she was a spiteful, selfish, lying girl who had spoiled everyone's fun by breaking up the fete early, and kept her indoors for a week.

–

Sophie woke with a start, drenched in sweat. She had not thought about that for twenty years, wasn't even quite sure it had actually happened. What had made her dream about it now?

She reached for the bedside lamp, her hand knocking something over. In the pool of yellow light, she saw it was the corn dolly, which had been propped up against the lamp. When had she brought that into her bedroom? She had left it on the mantelpiece, she could have sworn. On the bedside table was her wine glass and the empty second bottle. Sophie picked up the corn dolly and lay back on her pillows, holding it in both hands, wondering what had possessed her to buy it. It wasn't exactly ugly, just... its blank face and dry, rough limbs reminded her of that doll the police lady had shown her.

She dozed off again just as the pale dawn was insinuating itself between her curtains, and dreamed about the ghost asking her for a kiss, over and over and over.

9

Inside

Days in Withered Hill: 1

That first day in Withered Hill, Sophie runs out of the council chamber and down the stairs and out of the door into the blinding sunlight that bathes the market square, looking around to try to get her bearings and do what every nerve in her body is telling her to: get out of the village.

A woman pushing a pram and holding onto a toddler's hand pauses halfway across the flags to stare at her, and on the left side of the square is a small cafe, with tables outside on which a group of old men are gathered around mugs of tea, watching her with interest. Then, back on the cobbled Main Street, Sophie sees a policeman.

Some hidden memory inside her propels her towards him as he pushes his bike up the hill in the direction of the Post Office. She hears a door bang inside the council building and sets off at a run, waving and shouting at the policeman, who stops and turns to peer at her, shading his eyes from the sun.

He smiles at Sophie as she runs up to him, breathless and panting. He is in his forties and thick around the middle, and he pushes back his helmet to reveal a ring of sweat on his forehead, his dark hair plastered down. 'Everything all right, miss?'

'No, it isn't,' she says when she gets her breath back. She glances over her shoulder at the council hall. 'I… I think I'm being held here against my will.'

'Hmmm,' murmurs the policeman. He introduces himself as Constable Parry and says, 'You *think* you are being held against your will?' He makes a show of looking around. 'By who?'

'I can't really remember anything before I came here,' says Sophie. 'Please, can you help me get out of here?'

'Withered Hill? Leave Withered Hill?'

Sophie nods and looks back. The seven parish council members are now standing on the steps, watching her across the square. 'Please. Look. They're here.'

To her dismay, Constable Parry waves at the group, who wave back. What did she really expect? They're all in it together. Whatever *it* is.

The policeman smiles at her. 'It's natural you're a bit discombobulated at first. Always happens. You'll settle in.'

Carol and the man who seemed to be in charge – Noah something? – are walking in a leisurely fashion across the square towards them.

Constable Parry says, 'It's the way of things, see. You'll be all right. Just let them take care of you.'

'So I can't leave,' says Sophie numbly.

'Oh, I daresay you can. When you're ready to.'

'But when will that be?'

As Carol and Noah gently but firmly each take hold of one of Sophie's bare arms, Constable Parry says, 'It'll take as long as it takes.' Then he starts to whistle and continues to push his bike up the cobbled hill.

After leaving Mr Obermann at the library, Sophie decides to walk over to Catherine's cottage on the south side of the village. It's gone four and she should be home from the school. Perhaps she'll have some food on the go. Sophie can't be bothered cooking today. When she gets there and lets herself in through the red front door, hung with a wheat wreath – nobody in Withered Hill ever locks their doors – she realises that Catherine has got something other than eating on her mind.

Sophie wanders into the small living room and sits on the floral sofa, picking up a glossy magazine and flicking through it. Above her, in Catherine's room, the springs of the big metal-framed bed are squeaking rhythmically, and Catherine is crying out in increasingly frenzied, formless shouts in time with the banging of the bedposts against the wall.

The cries reach a crescendo and Catherine wails and lets loose a stream of swear words and invective that makes Sophie blush a little. Five minutes later, she comes downstairs, wrapping a silk robe around her, and grins broadly when she sees Sophie.

'Couldn't wait for Faunalia. Thought I'd give young Zeke Ellison a test drive.'

'And?' says Sophie, tossing down the magazine. There's a big photo feature on London in it, and it makes her feel both excited and nauseous at the same time.

'What he lacks in finesse he makes up for in…' Catherine holds her hands apart, like a fisherman boasting about his catch. She goes to the sideboard to get her cigarettes and lighter. 'He'll be ready again soon, I reckon. You know what young bucks like him are like. Go give him a ride if you fancy.'

Sophie wrinkles her nose as Catherine lights up a cigarette. 'I'm good.'

'Suit yourself. I'm going to go up for another go, then. Absolutely sure you don't want to join us?'

'Have you got any food?' says Sophie, shaking her head. 'I'm famished.'

'Cheese in the fridge. Bit of last night's hotpot on the stove. Help yourself.'

Sophie warms herself a bowl of Catherine's hotpot in the microwave for a couple of minutes and wolfs it down with a spoon just as the bed springs start bouncing again above her, and then she lets herself out of Catherine's cottage and heads towards the woods.

The woods grow in a thick ring all around Withered Hill, forming a green arch over the road in and out of the village. They are predominantly sessile oaks, their trunks gnarled and bent and spattered with lichen. These are mixed in with slender aspen and their dangling catkins, field maples, whose golden leaves elbow out the greenery come autumn, sycamore, poplar and horse chestnut. Sophie had been taught all their names not long after arriving in Withered Hill, learned how to recognise them just by laying a hand on their trunks. The meadows that buffer Withered Hill from the woods give way to thick undergrowth of bilberry, wych elm and purging buckthorn.

Sophie loves the way the spring sunshine filters through the canopy of the tall trees, dappling the ground with slivers of gold as she pads silently through the undergrowth, her pumps in her hand, her bare feet feeling their way along the cool earth.

She knows how to walk through the woods now. You need permission to be there, which is not obtained through asking. It is granted for purity of thought and stoutness of heart, for honesty and peace. You can never be part of the woods, never be *of* the woods, but you can be in the woods, for a brief moment in time, like a lover inside another. It is a bargain, a compact, a tryst. To partake of what it offers, you must leave something of yourself behind. Few people know that, thinks Sophie, as her hand trails over the wych elm in the shadows of the trees. It can be a memory of childhood, or a trinket given by a long-forgotten suitor. A baby's shoe, or a cat's claw. A secret whispered into a hole in the ground and covered over with leaves. A drop of blood.

Sophie unbuttons her shorts and lets them fall, stepping out of them and bending to pick them up, then walks on into the darkening woods. Without breaking stride, she pulls her vest over her head, then slides off her knickers. When she comes to a clearing, she folds her clothes on top of her pumps. Now they are off her body, the fabrics feel cheap and unnatural. Mass-produced items with no connection to anything meaningful. Designed to be worn and then fall apart and be discarded. But this is what she wears in the outside world. She leaves the pile of clothes at the base of an oak, trailing her fingers across its rough, lined bark, then walks on to the centre of the little glade, where a shaft of sunlight pools on soft, springy grass.

She has no fear here in the woods, not her. Not in Withered Hill's woods. She knows that dark eyes are upon her, but they will never hurt her, those folk behind those eyes. Never take what is not given freely. Sophie lies down on the grass, the sunlight painting her naked body. She thinks about Catherine in her bed with young Zeke Ellison, but there is no pang of jealousy. Not here. Here, she can breathe and think more clearly, shorn of doubts and worries just as she has shucked off her clothes. Grasses tickle her calves and thighs, and she feels an exploratory bramble stroking the inside of her wrist.

She has come here for guidance, for some inspiration of what she must do. The bower. Is this her bower? A lady's chamber, in the heart of the woods?

The bramble snakes around her wrist and she closes her eyes. To take from the woods, she must also give to it. She lays her other arm flat on the ground, palm up, and allows another writhing bramble to curl around her wrist, its spikes digging into her. It hurts, but in an exquisite way. It feels as though the grasses are growing around her, over her, gently but firmly claiming her, at least for a short time. She submits to the woods and the hungry eyes that watch her, and opens her mind to allow whatever wisdom the trees might offer to flood in. And in return she will, eventually, as the grasses and roots and brambles range over her body, parting her legs, holding her wrists, cry out just as Catherine cried out in her metal bed, and that is the thing she will leave the woods in return for whatever it might gift her.

Days in Withered Hill: 1

When Constable Parry hands Sophie back to what she can only think of as the custody of Carol Mountjoy and Noah Jones, they take her to the cottage of Catherine Baldwin, where she will stay until she is, as Carol says, 'more settled'. By which, Sophie thinks, she means until she stops trying to run away.

Catherine is sitting at her kitchen table when they let themselves in, smoking and staring at Sophie with a look she cannot read. The blonde woman is wearing a tight shirt that is unbuttoned enough to show off her cleavage, and a pair of jeans.

'So this is her, then.' A statement, not a question.

'Sophie, meet Catherine Baldwin. She's a teacher at our local school. You'll be staying here for a few days. Until you get your bearings,' says Carol.

Sophie looks around the small cottage. 'So I'm a prisoner here?'

Catherine exhales smoke. 'You could leave today, if it was up to me. Can't say I'm turning cartwheels at the thought of you being here.'

'Well,' says Carol, breaking the awkward silence. 'We'll be off, then. Let you two girls get to know each other.'

When they've gone, Sophie sits down warily at the table, the half-full cut-glass ashtray between them. Catherine says, 'Do you smoke?'

'No,' replies Sophie, watching her stub out her butt in the ashtray. 'I don't know. I don't think so.'

'Well, if you decide you do, don't be smoking all mine. Price of the damned things.' Catherine puts her head on one side and looks at her for a long moment. 'You're quite pretty. Nice long legs, too. Good tits.'

Sophie looks down at herself self-consciously. 'Um. Thank you.'

Catherine lights up another cigarette. 'It'll be handy when you get to leave. Girls who look like you always do well in life. Unconscious bias. So what do you actually know about yourself?'

Sophie shrugs.

'Sophie Wickham, thirty... one? Two? I can't remember,' says Catherine. 'Born in a small village near Gloucester I can't recall the name of. Dead-end job doing something pointless. Used to have a boyfriend called

Jamie. Except he was a dick. Parents both dead. Car accident.'

Sophie gapes at her. 'How do you know all this?'

'Social media, mainly,' remarks Catherine. 'It'll all come back to you, I'm sure.'

'A car crash,' says Sophie softly. She tries to conjure up an image of her parents, but she can't. All she can see is a car slamming into something indistinct in slow motion, two faceless crash-test dummies in the front seats. How come she knows what a crash-test dummy is but can't remember her own mother and father?

'Look, this is never easy,' says Catherine, her tone softening slightly. She stands up and goes to the fridge. 'Do you like wine?'

'I don't know,' replies Sophie miserably. 'I don't know what I like or what I don't like.'

'Well, maybe we should find out then,' says Catherine, pulling a bottle from the rack in the fridge door and getting two glasses from the cupboard above. 'If I get drunk, don't let me forget to put a bowl of milk outside the front door before I go to bed.'

'What for?' asks Sophie.

Catherine laughs as she pours two big glasses of wine. 'You really do have a lot to learn about Withered Hill.'

10

Inside

Days in Withered Hill: 1

Dusk falls heavy and thick on Sophie's first night in Withered Hill, Catherine's kitchen window darkening to a black mirror that looks out onto the hills and meadows to the south of the village. Catherine has smoked half a packet of cigarettes and gone through the best part of two bottles of white wine while Sophie still sips at her glass. She doesn't think Catherine has noticed she isn't really drinking.

'And tha'ss why he was a cunt,' slurs Catherine. She has been giving Sophie a rundown of her colourful love life with men both from Withered Hill and elsewhere, none of whom she seems to have parted with on good terms. Sophie has lost track of which one they are talking about now. 'Need a wee,' says Catherine, lurching unsteadily to her feet and edging along the table until she can make a grab for the kitchen door.

When Sophie hears her stumbling up the stairs, she pours her wine into Catherine's glass and refills hers with water from the tap, as she has been doing for the past two hours.

The toilet flushes noisily above her and Catherine pounds heavily down the stairs and back into the kitchen.

She sits down and drinks deeply of her wine, then looks at her watch. 'Goodness. The time. Up for school tomorrow. Bed. You're in the spare room. I made it up for you.'

'You told me to remind you something about milk…?'

Catherine taps the side of her nose with an exaggerated motion. 'You don't really need to remind me. Nobody ever forgets that.' She goes to the cupboard to get a dish and then fills it with milk from a glass bottle in the fridge. 'Get the door for me.'

Sophie walks ahead of Catherine in the narrow corridor, opening the wooden front door that's just on a latch so she can lay the dish down on the step, milk slopping over the rim.

'What happens if you don't do it?'

'Bad,' says Catherine, standing up. 'Very, very bad.'

Sophie feels the closeness of Catherine's body as they stand in the doorway. She smells of cigarettes and lavender and something else, something musky and dark and somehow forbidden. Catherine is a head shorter than Sophie, and she turns her face up to her.

'Do you want to kiss me?' murmurs Catherine. 'Everybody wants to kiss me.'

'I don't know,' says Sophie truthfully.

Catherine sways a little and holds up a finger in front of her face, waving it from side to side. 'Well, don't want that. It won't get you anywhere. Especially not out of Withered Hill. Bed. Now. Up early tomorrow.'

Catherine has laid out some cotton pyjama bottoms and a black vest on top of a single bed with a patchwork bedspread in a small room at the back of the house. The other woman staggers into her room next door and Sophie hears her start to snore gently within minutes.

Sophie lies on the bed, still in her dress and pumps, clutching the pyjamas to her chest, and stares at the ceiling, and waits. When she is sure enough time has passed, she steals down the wooden staircase, painstakingly lifts the latch on the front door as quietly as possible, and slips outside into Withered Hill's black, moonless night.

Catherine's cottage sits apart from the main clutch of houses on the south side of Withered Hill, slightly askew and facing towards the woods, which rear up across the meadow as a darker mass than the night, like a jagged hole, or an absence of light. Sophie pads lightly through the tall grasses and spring flowers towards the trees. She has no real idea of what she will do when she gets out of Withered Hill, but deep in her gut, she knows she does not belong here. She still cannot remember anything at all of where she has come from or what she did before she arrived this morning, but she trusts that somehow, somewhere, she has a life outside the village, and all she has to do is find it again.

There is no path through the trees, and at ground level there are thick shrubs and masses of brambles, which Sophie picks her way through carefully, flattening them down with the soles of her pumps. Even so, she feels them scratching at her bare calves, and she is relieved when the bushes thin out and give way to soft grasses beneath the trees.

She walks with her arms outstretched as she gradually gets used to the pitch darkness, and after a while she can make out the shapes of the different trees, though she has no names for them yet. How deep are the woods? She tries to recall from when she stood on the hill by the Post Office and could see the ring of trees around the village. And she came in through the woods; she must have. She wishes she could remember. Wishes she could remember anything at all before Withered Hill.

Something flaps and screeches and takes flight close by her, crashing through the low branches of the trees, making Sophie cry out. A bird. An owl, or maybe a wood pigeon. It vaguely annoys her that she has names for these things but cannot remember anything important.

Something scratches her leg again and stings sharply. She has wandered into more brambles. Does that mean that she is coming out of the woods already? The trees do seem to be thinning and she picks up the pace.

Yes, there's a definite clearing ahead, she thinks, as she pushes on through the thickets. Beyond is blackness, but not the packed blackness of trees. Open land.

Sophie stops at the treeline, looking out in dismay. She is back where she started, staring across the meadow at the dark shape of Catherine's cottage, and the quiet houses beyond. She swears quietly. She must have got turned around in the woods.

Taking a deep breath, she spins on her feet and pushes back towards the trees. The brambles rip at the hem of her dress, and she bends down to tear a strip of the light cotton from it, tying it around a low branch of a tree a couple of hundred yards into the woods. If she sees it, she'll know she's lost her way again and been turned round.

Sophie starts to count her steps, making sure she is walking in as straight a line as possible. Ten. Twenty. Thirty. A hundred. Two hundred. Surely the woods aren't that deep. Something flutters by her face, a moth or spider's web, and when she reaches up to brush it away, it is the strip of her torn dress. She is back where she started.

Three more times Sophie attempts to push deep into the woods, three more times getting turned back, and as

75

she stands, crying with frustration and anger staring at Catherine's cottage, she can see the thin light of dawn threatening the eastern sky. How is this possible? How can she not find her way through a copse of trees at the edge of a village?

She bellows a formless cry of fury and turns and runs, head down, into the woods, ignoring the twigs that strike her face and the thorns that rip her dress, the brambles that seem to entwine around her ankles, raising bloody welts as she kicks them away.

Grey light is filtering through the thick canopy of the trees and she thinks, *This time,* this time*, I will not be turned around, I will keep my focus, this time I will do it. I will leave Withered Hill.*

And then she sees them.

They are as though carved from the bark, figures bound like twigs with twine, but huge, sometimes as big as a man, sometimes smaller, sometimes with far more dark, gathered mass. They are feathers and fur and beaks and teeth and yellow, raptor eyes. They are skin and bone and rags and shadows, sprouting from the land and hanging from the trees. They are the fruit of the woods and the heartbeat of the earth and the devils in the details of the rough, hard-hewn land on which Sophie stands like an alien entity that does not belong there. They chitter and chatter and sing like birds, squeal like mice, grunt like pigs, and together they are the most terrifying thing Sophie hopes ever to see in her life.

They seem at once far away and then suddenly all around her. Keeping their distance, as though cautiously, and then poking and prodding her, plucking at her dress. They laugh and squeal and make raw, guttural noises that speak of hunger and desire.

Sophie does not know if they are the men of the village, in their animal heads, blood flowing down their necks and chests, emboldened by the darkness to do more than look at her and point the way she should walk, or if they are something else.

She has an overbearing feeling that she belongs to them in some way, as they flit among the trees and mutter and grumble and touch her legs and tug her hair. They whisper her name, like the buzzing of bees or the chirping of crickets. *Sophie Wickham Sophie Wickham Sophie Wickham.* Their knowledge of her name gives them power over her, more power than she has over herself. They know more of her than she knows, and that is the claim they stake on her.

Sophie turns and flees, back the way she came, ignoring the grasping branches and the tearing thorns, and by the time she crosses the brambles, her dress has been ripped from her in tiny shreds and her naked body is a skein of cuts that bleed and sting. She collapses in the meadow as the sun casts its first warm beams across her, and sobs into the dirt, waiting for someone to come and take her home.

Days in Withered Hill: 358

They are watching her now, thinks Sophie as she dresses by the oak tree, wriggling into her shorts and pulling her vest over her head. They would not have harmed her on that first night, though she understands full well how she feared they might. She was trying to leave Withered Hill before her time. They were merely not allowing that to happen. Just doing their job.

She stands for a moment in the quiet solitude, watching mayflies dance in the lances of sunlight. Beyond the clearing is a swampy wet stretch where frogs croak at dusk. It's the most beautiful spot Sophie can imagine. She has lain here on the land and given of herself, and if the land has given her anything in return, she is not yet aware of it. But she knows it doesn't necessarily work like that. Sometimes the message takes time, sometimes it comes in a less obvious form.

Sophie thanks the earth for its ministrations and walks back through the woods towards the village. She pauses at the tree where the scrap of her dress hangs, faded and tattered, but still tied to the branch where she left it that first night. Back when she thought Withered Hill was a prison.

She touches it lightly, and then walks on and out of the woods, into the meadow beyond which Catherine's cottage sits. The door opens and Zeke Ellison lets himself out, looking shell-shocked and exhausted. Sophie laughs to herself, and Catherine stands at the door, miming putting a cup to her lips. Sophie shakes her head and points towards the village, to her own home.

She takes one last look at the woods, feeling the weight of those many eyes upon her and wonders what wisdom, if any, they have imparted during her sojourn among the trees.

11

Outside

Days to Withered Hill: 9

Sophie woke up to the tension-evaporating relief of payday, and the gut-wrenching realisation she had a date that night. She immediately got on to her banking app and squirrelled money into two or three different online accounts she had; it would stop the people she owed it to grabbing it straight away, and though there'd be a barrage of phone calls over the rest of the week, at least she had breathing space today.

Sophie showered, washing away the detritus of sleep. She'd dreamed of the hare in the swaddling clothes and the nursery again. And there had been other dreams. They had been dark and troubling, mainly about Tom Gisburn. There had been something... animalistic about him. Wolfish. No, that wasn't right. Fox-like, perhaps. Cunning, sly. He'd reminded her of Jamie, in some ways, and she realised she was just mapping onto Tom her anxieties, because her last relationship had hardly been sunshine and flowers and hadn't ended in a way that anyone would call healthy. Still, she hadn't heard from Jamie for ages and his socials had been quiet since about the same time. Though he'd blocked her on every platform after they broke up, and then kept sending her repeated friend requests for months afterwards, she could see there hadn't been much activity, if any. Maybe he'd just left the country and gone to live up a remote mountain somewhere. She could only hope.

Tom Gisburn was not Jamie, she told herself sharply as she wrapped a towel around her hair, then cleaned her teeth. She spat into the sink and was brought up short by the sight of blood; not a lot, and there was something in the foam of red-tinged toothpaste. She peered closer; it was a tiny strand of dried corn stalk, obviously lodged in her teeth and pricking her gum. What had she been doing to that corn dolly in her sleep? Eating it?

In the kitchen, Sophie grabbed a pair of leggings from the clothes dryer and squealed as something scuttled on the periphery of her vision. She'd have to tell the landlord the cockroaches were back. It sat there against the skirting board, near the fridge. Backing slowly away, she grabbed one of her boots, sidled up to it as its antennae twitched, and brought the heel down hard, feeling the satisfying, squelchy crunch. You weren't supposed to do that, she'd read somewhere. In their death throes, they somehow sent out a message to all the other cockroaches, which brought them running. She imagined a silent, psychic cry for help, and the other insects gathering around their fallen comrade, heads bowed in prayer. Wincing, Sophie gathered the crushed remains up in a wad of kitchen roll and threw them in the bin.

While waiting for the kettle to boil, she noticed the corn dolly sitting propped up against the toaster. Had she got up in the night for a glass of water and left it there? It was holding Tom Gisburn's business card; or, rather, the card was neatly stuck to the dolly's chest with a sewing needle. Sophie decided to take her coffee black and, not for the first time, felt the dull pain that clung to her head like an aura was tinged with hues of vague shame. She really should try to stop drinking so much.

From the corner of her eye, she caught a movement: another cockroach, antennae twirling like whiskers, investigating the smear on the linoleum where she'd dispatched the last one. So the story was true, then. She got her boot and it suffered the same fate. No doubt there'd be another one there when she got home from work. They were relentless, creepy little bastards. Just like Jamie.

She had met him online, getting caught up in a couple of comment threads about TV shows and movies, then one morning woke up to find he'd liked dozens of her posts in the small hours. And not just her selfies; her carefully composed thoughts on politics and celebrities and news events, too. Being a woman on the internet was a frustrating business at best. A pithy tweet got a handful of likes, a photo of her in the pub earned a volley of friend requests and love hearts, sometimes private messages along the lines of *Hey babe love your page.* Jamie was different. He followed her, she followed back. They had a nice, easy back-and-forth on each other's timelines. They liked a lot of the same things. He read books and recommended stuff to her and was interested in what she thought about things. They increasingly had their conversations in private and one Sunday morning she woke crick-necked on the sofa, two empty wine bottles on the rug, her face burning as she read back the conversation she had no memory of. Sexy stuff. Detailed descriptions of what they'd do to each other. She was shocked at her own uninhibited, flowing prose, quietly impressed by his masterful claiming of her with words. Her cheeks burned as she read her own increasingly shambolic typing that described her orgasm.

As she was reading it all back, he sent a tentative *Hi*. They chatted about inconsequential things, the thread of their lust hanging above the mundane conversation. By ten o'clock that night, they were at it again, Jamie indulging Sophie's most secret fantasies that she'd never disclosed to anyone before.

He was in London. They met for drinks, and sex. They went to the cinema, book readings, walks in the park. They morphed into boyfriend and girlfriend without anyone actually saying anything.

Sophie had never been out with someone so thoughtful. She would frequently get flower deliveries at the weekends. He would hunt down first editions of her favourite books, and buy her expensive jewellery that he would hide in boxes of cheap chocolates. On the anniversary of her parents' death, he drove her to her old village and booked them into a huge country house hotel in the Cotswolds. In the morning, he presented her with a sapphire necklace that took her breath away. It was exactly like the one her mother had been wearing when she died, and which, Sophie had told Jamie months before, her mother had promised her, but it had been burned up and destroyed in the car crash along with both of her parents.

At first, she found his slight jealousy endearing. He would go into a sulk if she interacted with another man online. He would trawl through her profile and point out casually that she was liking a lot of posts by one guy or another. If she didn't message him by eight or nine o'clock at night, when they eventually spoke he'd drop into conversation that it was fine, she must just have been chatting with someone else. Increasingly, Sophie found herself having to salve his hurt feelings, reassure him that she loved him, strenuously deny that she was flirting with other men. She closed her direct messages so only friends could send them, and she stopped posting as many selfies. Her thumbs would hesitate over the phone keypad before typing a contribution to an online conversation, and invariably she would just not bother. Jamie never overtly told her to do or not do any of these things, but somehow she just gradually and subconsciously came to the conclusion that it all made for a quieter life.

So, of course, when she eventually did betray him, his passive-aggressive behaviour boiled over into something that was a very long way from passive.

—

Sophie's job for the day was typing up a handwritten technical report about, as far as she could gather, soil erosion on the banks of several rivers in Teesside. Mid-morning, Colin brought her a coffee from the machine that she hadn't asked for, but which she graciously accepted.

'Any plans for the weekend?' he said, his words dripping with tragic hope.

'The weekend starts tonight, actually,' she replied, not taking her eyes off the report, her fingers dancing over the keyboard. 'I've got a date.'

'Oh,' said Colin, sounding slightly crestfallen.

Mandy was passing her desk and glared at the both of them. 'Is this an officially sanctioned break, Colin? And for *anyone* who has a date tonight, might I remind you that we start at nine promptly every morning.'

When Mandy had bustled on, Colin said, 'Anyone nice?'

'I hope so,' replied Sophie, feeling a sudden, slight affinity with Colin against Mandy's tyrannical rule. 'It's our first date. I only met him this week. By accident.'

She looked up after Colin fell silent, seeing him frown. 'By accident?'

Sophie smiled. 'He walked straight into me. Spilled coffee down my front. Monday. Do you remember what I looked like when I came in?'

'You always look lovely.'

Sophie felt herself icing up towards Colin again. Why couldn't men just have a fucking conversation without bringing it back to how a woman looked? She should have known better than to try to be nice to him.

'What's his name?'

Sophie started to say something, then bit her lip. 'You won't know him.'

'Where's he from?'

Sophie looked at him, annoyed. 'Why?'

He looked like he wanted to say more, but Mandy was on her way back. She glared at Colin and tapped her wristwatch with a thick forefinger. Then she looked at Sophie. 'Can I speak to you in my office?'

Fifteen minutes later, Sophie emerged feeling slightly nonplussed. She had thought she was going to be fired. It turned out the company was offering her a full-time contract. Sick pay and holiday pay and regular shifts. 'I thought you didn't like me?' Sophie had blurted out.

Mandy had looked at her with veiled eyes. 'I don't like *anybody*. But your work is good and we're impressed with your output, Sophie. Sign here, here and here please.'

Colin was crouched down by her desk, dabbing at the carpet tiles with a wad of tissue paper. He looked up at her mournfully. 'I was looking for a stapler and I knocked over the coffee cup. There wasn't much in it. I'm sorry.'

Sophie sighed and opened her drawer, handing him the stapler. He stayed on his hands and knees, looking up at her like a puppy. 'Can we all get on?'

She sighed.

Colin nodded and scurried back to his desk.

Sophie worked for another hour, then broke for lunch. For the first time since she'd worked there, the clock didn't start counting down on her monitor. She was a full-time, regular employee now. And that obviously came with some benefits.

It was another nice day, so Sophie treated herself to a Pret sandwich and coffee and sat on the bench outside the church, watching the people hurrying along the pavements. She wondered where Tom Gisburn had been going to, coming from, when he'd walked into her. She glanced at her phone. She'd be finding out, perhaps, in eight hours.

Sophie threw her sandwich wrapper in the bin and headed back towards the office. She paused before crossing the road; at the little cobblers and key-cutting shop wedged between a Boots and the bank, she could see Colin, handing over his card in exchange for a brown paper bag. Sophie didn't want to have to walk in with him, so she meandered along the kerb, waiting until he hurried back to the revolving doors, then crossed over and went back to work, and to mark time until the date that was making her insides feel all liquidy with nervous excitement.

12

Outside

Days to Withered Hill: 9

Halfway through the afternoon, Colin came over with another unasked-for coffee, clumsily kicking her bag over and spilling the contents across the floor: her phone, purse, tampons and, to her surprise, the corn dolly. Sophie didn't remember bringing that to work. Those cockroaches must have freaked her out more than she realised. Colin started to mutter apologies and fell to his knees, gathering up her stuff.

'It's fine,' she said, taking her keys and phone out of his hands and stuffing them into her bag. 'I'll do it.'

Colin was holding a packet of condoms and looking at them like he didn't quite know what they were. Sophie snatched them from him. They'd been in there so long they'd probably perished. She wondered if she should pick some more up for tonight, then reprimanded herself inwardly. She'd already promised herself that she wouldn't sleep with Tom on the first date.

'That's nice,' said Colin, pointing at the corn dolly, but not moving to pick it up. 'What is it?'

Sophie shrugged as she shoved it in her bag. 'Just something I bought. I liked the look of it.' Her voice dropped to a mutter. 'Apparently.'

'People used to use them for protection.'

Sophie glanced at Colin with narrow eyes. He knew a lot about it, suddenly, when three seconds ago he was asking her what it was. 'Protection from what?'

'Dark forces, I suppose,' he said, getting inelegantly up to his feet. He brushed his fringe out of his eyes. 'Lot of old nonsense.'

So why bring it up?

Sophie turned back to her screen, but Colin continued to lurk, like an odour. 'Going anywhere nice on your date?'

'Yes,' she said without turning to him.

'Well, don't do anything I wouldn't do.' He gave an odd, high-pitched laugh that was singularly unpleasant.

Why wouldn't this creep just leave her alone? She swivelled towards him in her chair and fixed him with her gaze. 'Oh, I imagine I'll be doing things you haven't even dreamed of in your darkest moments, Colin.'

He held her stare for a moment and then his cheeks flushed and he looked away. He smiled awkwardly and shuffled back to his desk.

Sophie carried on with her work, and thought about those condoms in her bag, and tried to ignore the insistent chattering between her legs.

–

Sophie suspected that Jamie had at some point read a book called something like *How To Be A Considerate Lover*. He was gentle and loving and kind and constantly murmured to her about how her pleasure was so much more important to him than his. His adventurousness in their early flirty text sessions hadn't translated to real-life passion. At first, she thought it sweet, but it quickly began to pale, and it started to irritate and frustrate her. If it was indeed true that her pleasure was more important to him than his own, then it felt like he was silently waiting for a medal for it, and besides, wouldn't it have been a good idea to actually find out what her pleasure was in the first place?

One night after the pub, as they lay in bed and Jamie prepared to run through what Sophie by then felt was like a mental checklist, she threw her arms back over her head and fixed him with her very best wanton stare. She crossed her wrists on the pillow above her and whispered, 'Hold me down. Call me a slut.'

Jamie stared at her, then moved to his side of the bed, turning his back to her and sullenly pulling the duvet up to his neck. 'I'd never say that to you. Because you're not. Why would you even want that?'

'I know I'm not,' she said, putting a hand on his shoulder, over the duvet. She thought about it for a moment. 'Or maybe I am. Sometimes. Just for brief moments. When I'm with you. Nobody has to be just one thing all the time, do they?'

'You do,' said Jamie, shrugging her hand off. 'You're beautiful and perfect and wonderful. Why wouldn't you want to be that all the time?'

They lay in silence together for a while, sleep eluding Sophie, though she was a bit tipsy. When Jamie started gently snoring, she quietly picked up her phone and dimmed the screen, scrolling through her socials. She had noticed a week before that Jamie had followed, or sent friend requests, to every single person she was connected with on all platforms. She couldn't do anything without being under his watchful, protective, loving gaze. Two days previously, feeling good about herself, she had posted a selfie in a new summer dress, the light from her bedroom window shafting in and picking out the silhouette of her body beneath the thin fabric. It got a lot of likes. Of course, Jamie was the first to pointedly like it, and the first to comment, with a proprietorial *Looking fab, babes. Glad you're mine. Love you xxx*.

Idly, she scrolled through some of the other likes. Somebody had reposted her, which meant the picture had gone into the timelines of people who didn't follow her. A profile caught her eye, with a black and white photo of a man in a suit, shirt open a couple of buttons to reveal a hairy chest, the head and face cut off. His bio simply said *You can call me Sir*. He posted a lot of arty monochrome faceless selfies, sometimes in his tight-fitting underwear. Lots of pithy soundbites about giving a woman what she needed by telling her what she wanted. Her hand trailing to her inner thigh under the covers, Sophie glanced at Jamie and then began to like almost every post.

Within two minutes, a direct message notification popped up. From him. @CallMeSir. It said, *Good evening.*

Sophie bit her lip, looked at the snuffling Jamie one more time, then opened up the reply box.

At the time, Sophie was temping in an insurance office with a small team that was almost exclusively female. Jamie fully approved of this. But when Sophie told him she was having a night out with her workmates, his face fell and he slumped into a sulk, making comments about what girls were like when they all went out together and got drunk, and what men were like when they saw girls like that, and how he'd seen it happen and men could be such shits and he really hoped she'd be all right, and that she should check in with him every hour or so, and he could come and pick her up if she was having a bad time.

'I won't have a bad time,' she said, kissing him on the nose, and letting him take her to bed and put her pleasure above his own, pleasure she faked enthusiastically while telling him no man could ever do for her what he did.

Of course, there was no girls' night out. She had been talking to @CallMeSir at every available moment, sometimes even when Jamie dozed on the sofa. Their exchanges had been heated and unfettered and exciting. He had sent her pictures of his face, which was rugged and bearded, and he told her his name was Phil. After two weeks, they had agreed to meet for a drink, at the other side of London.

Phil was indeed handsome and fit and refined and looked at her with restrained hunger as they had a drink in a bar that played smooth jazz. He was also screamingly narcissistic, talked about himself all evening, showed her photos of himself on his phone she might have missed on his timeline, asked her almost nothing about herself, and evidently fully expected that she would have sex with him. He was also married but hadn't 'had relations' with his wife for two years, and he was only staying because the kids were so young. Sophie was utterly turned off and declined his offer of a hotel room – though he wouldn't be able to stay the whole night, of course – unfriending and blocking him on the night bus home, and thinking that perhaps Jamie wasn't so bad after all.

When she let herself into her flat, Jamie was waiting for her on the sofa, sitting in darkness, save for the light from his phone. She walked over to him and he held it up wordlessly. He was logged into her account, and his screen displayed her messages with @CallMeSir.

'How did you get my password?' she said.

'It was easy to guess,' said Jamie levelly. 'You're not actually that smart, Sophie. Anyway, that isn't the issue here.'

Actually, it did somewhat feel like the issue here. But her cheeks burned as the evidence of her infidelity glowed from the phone screen. 'Nothing happened tonight,' she said numbly. 'I didn't like him. You're angry. I get it. I'm sorry.'

'I'm not angry, I'm disappointed,' said Jamie, as though talking to a child. He nodded to her own phone in her hand. 'Now, I think it would be best if you deleted all your social media accounts, don't you? You don't need them. You've got me.'

Sophie stared at her phone and started to cry. 'All of them?'

Jamie stood and faced her. 'All of them. It's just social media. It means nothing. You don't need to post pictures of yourself for likes. I'll tell you you're beautiful. If you've got something funny to say, say it to me. I'll laugh.' He looked at his phone. 'If you want to have sex, then I'll have sex with you. And I'll be more considerate than this prick.' He looked at her. 'You can't really like all these horrible things he was saying he wanted to do to you. You're not like that.'

Sophie opened up the account settings on her screen. Her thumb hovered over the deactivate button. She looked at Jamie. 'No.'

He half-smiled. 'No? You mean, *no, I'm not like that?*'

'I mean, no. I won't do it.'

Jamie tugged at his chin, as though lost in thought, frowning as if he just didn't understand. 'No?'

'No,' said Sophie firmly.

She didn't know he could move so fast. The back of his hand connected with her cheek and her head snapped back, and she lost her footing and landed heavily on her backside on the rug.

'Is that what you want?' Jamie screamed. 'Is that what you like? A bit of rough? Do you want me to hit you again? Is that what gets you off?'

'Get out,' said Sophie calmly, her hand on her burning cheek. 'Get out or I'll call the police.'

Jamie's eyes filled with tears. 'Oh God. I'm sorry. It's just… I don't know what you want, any more. I don't know who you are. I don't know what you want me to be.'

'I want you gone,' said Sophie.

'Please,' he said, suddenly sobbing.

'Now.'

Jamie nodded and walked to the door. 'I'll be back tomorrow. We'll sort all this out then. Don't worry. We'll get through this. Just so you know, I forgive you.'

When he'd gone, Sophie ran to the door and bolted it. She texted the landlord, though she knew it was late and he'd be in bed. *I need the locks changing on my flat as soon as possible.* Then she went and opened a bottle of wine and drank herself, sobbing with relief and release, into unconsciousness.

—

Sophie sprayed herself liberally with Chanel No. 5 and slipped into the tropical print dress, then put on the cork wedges, and found her dark green handbag. She'd given her hair a bit of a loose, bouncy curl. She did, she had to admit, look good. Better than good. She finished the last of the glass of wine and then applied her lipstick.

She was nervous. Not all men had to be a Jamie or a Phil, she knew. She prayed that Tom Gisburn was neither. She thought about his eyes, the way he dressed, how strong his hand had felt when he'd helped her up off the pavement.

Everybody was different, really, weren't they? Maybe this was going to be good, after all. Part of her wished Jamie could see her now. He hadn't beaten her down. She could do this. She could rise again. Perhaps she could even be happy.

Sophie blew herself a kiss, and went to meet Tom Gisburn. But she had a stop to make on the way first.

13

Inside

Days in Withered Hill: 359

The next day is a Saturday, and Catherine calls round to Sophie's cottage to ask her if she wants anything from outside. They sit on the sofa that's draped with a candlewick spread, sipping tea. 'I could put a clothes order in for you?' says Catherine. 'Maybe some summer dresses?'

The internet does not work in Withered Hill, nor do mobile phones. Nobody has ever run cables there, nor built a mast nearby. Some people say it is the thick ring of trees, others simply say that *they* wouldn't like it. Withered Hill seems to get by without it, though, and Sophie has a good understanding of what it is, thanks mainly to Catherine's teachings. As well as educating her young charges at St Michael's, Catherine has also spent the best part of a year tutoring Sophie in the ways of life outside. Sometimes, the knowledge feels new, as though poured into her head. Other times, like it is a memory to be unlocked.

Catherine is one of the few residents of Withered Hill who leaves the village frequently, and when she does, she uses her phone, suddenly bursting with connectivity once beyond the parish bounds, to buy things online, or to download pages and pages of Sophie's old social media accounts to be printed out at the Post Office, for her to pore over, immersing herself in the London life.

Sometimes, Catherine buys clothes for Sophie direct from shops outside, or books, or trinkets. One time, just a few weeks before, at Beltane, she brought something else.

Days in Withered Hill: 319

'We call it Beltane because that's what they used to call it way back when,' says Noah Jones, spitting tobacco on the pavement. 'But really it's our own thing, like everything we do in Withered Hill.' He looks at her sidelong and lifts his cap off his forehead; there is a line of sweat there. It is the last day of April and there has been something of a mini heatwave for the last few days. 'I imagine tha'll find it interesting.' He pauses. 'Tha's not going to try to escape again, are you?'

It has been a full month since Sophie tried to flee Withered Hill. She hasn't exactly given up, but she is more… reconciled to the idea of not running away. In fact, she sometimes doesn't think about it for days on end, as immersed in village life as she has become. And besides, she is constantly given assurances that at some point she will leave Withered Hill. It's just that nobody knows quite when, and they won't tell her why they don't know, or who will eventually make this decision. And nobody has told her what she will do when she goes, where she will live, how she will survive. Her knowledge and memories are fragmentary, and she doesn't know what she's learned or what she is starting to remember. Withered Hill, for all she is a prisoner here, feels safe; will she feel the same out there?

'That were a good trick, mind.' Noah chuckles. 'Hiding in the back of the delivery man's van. Though tha' knows tha' wouldn't have got over the village bounds. He'd have had a blowout, or his engine would have seized, or he'd have remembered he'd not delivered a parcel somewhere.'

As it was, the courier had trundled up the main street and then stopped to double-check his manifest, which he'd accidentally left in the back of his big white van. And when he opened the doors, Sophie was crouched there among the boxes and packages, and he'd ordered her out just as Constable Parry came up, wheeling his bike, and smiled indulgently at the driver and told him that it was 'one of those Withered Hill things'.

The outside world came into Withered Hill infrequently, but it did come. And it had learned, somewhere along the way, to not question what went on in Withered Hill, or the ways of its people. Even the protestations of a pretty young woman that she was being held in the village against her will.

'No,' says Sophie, eventually. 'I won't try to escape, I don't think. I'm interested in seeing Beltane. What is it?'

They watch as the children from St Michael's rehearse their performance around the big maypole in the village square. They are to sing folk songs and act out a little play that Catherine has written about the founding of Withered Hill. Beyond, there is work being done: setting up a beer tent and building a bonfire, and hay bales are being placed as seating for the villagers.

'There's always been a celebration on May Day,' says Noah. 'We give thanks for the good fortune we've had in Withered Hill, and ask for our seeds and bulbs and animals to be blessed so they give us a good harvest. We are what we eat, in Withered Hill; or rather, we are what other folk eat. Our crops in the fields, our fruit in the orchards, our meat from our livestock. You can tell when you've eaten produce from Withered Hill, even if you don't know it.'

'A blessing from who, though?'

Noah smiles indulgently at Sophie, just like Catherine does with her pupils. 'Tha' knows, Sophie Wickham. Don't play silly buggers with me.' He squints, scanning along the road, the one that leads out of Withered Hill. 'Besides, tha's got the starring role this Beltane. Look, here comes Catherine's motor now.'

—

In the early evening, before the sun sinks over the distant green hills, everyone gathers in the village square for the Beltane celebrations. True to Noah's word, Sophie is given a prime spot on the front row of hay bales, sandwiched in between all the members of the parish council. There is a band made up of some of the regulars from The Farmer and Devil and with Catherine singing; Sophie has seen them play many times and is always transported by their hypnotic drums and fiddles and her friend's honeyed, silky voice singing tunes of the land and the sky and ways of life forgotten by the outside world. There is a huge pile of sacks and boxes of grain, seeds, bulbs and animal feeds, taller than Sophie, at the centre of the square, and not far from it a small wigwam of bonfire wood.

After the younger children have danced to the band around the maypole, the multicoloured tapes winding in on themselves as the children holding them skip in and out of each other, with an almost mathematical beauty, it is the turn of the older children and their play.

Lucy, the girl with red hair in pigtails, is the narrator, and she stands in a simple shift dress as the crowd quiets and the men from the pub, including the players in the band, shift roles to become scene dressers, wheeling big canvases behind the children as they perform. The first one shows the rolling moors around Withered Hill. A boy dressed in a smock and wearing a straw hat walks slowly across the scenery, a girl in tow behind him.

Lucy says loudly in a monotone delivery, 'In the very olden days, before Withered Hill was really a village at all, there were just a few farms in this area. This particular year, the harvest had not been good for anyone and the people were very worried. They had barely enough to eat, let alone to sell at market.'

The girl says, 'Father! What are we to do?'

'I do not know, Daughter!' says the boy, too loudly. 'Things are very bad for us, that is for sure!'

'But there was help at hand,' says Lucy. 'For the farmer and his daughter were about to meet someone very special.'

From behind the scenery, another boy, wrapped in a ragged black costume and with his face painted white, jumps out, and all the crowd applaud. Sophie looks around and haltingly claps too.

'Who are you?' yells the farmer.

The new boy hunches over and glances back off the makeshift stage-set. Sophie sees Catherine nodding encouragingly at him from the wings. He clears his throat and says, 'I am Owd Hob!'

'What are you?' asks the farmer. 'Boggart? Fairy? Sprite?'

Owd Hob looks back at Catherine, then says, 'All of these things and none of them. I am of the land and the land is of me.'

'Well, the land has not been very good to us this year!'

Owd Hob scratches his chin. 'I can do something about that. But you will have to do something for me.'

'I am just a poor farmer and I do not even have a good harvest! What would you have? My last turnip?'

The crowd laughs.

Owd Hob says, 'I would have a wife.' He peers around the farmer. 'Your daughter is a pretty one. Let me take her as my wife and I will promise you…' He looks back and Sophie sees Catherine mouthing words at him. '…I will promise you a bountiful harvest.'

'Father! I do not wish to be married to this Owd Hob!'

'And I do not wish you to marry him, Daughter!' The farmer stalks up and down, scratching his head. 'But something must be done!'

'I will give you until sunset tomorrow!' says Owd Hob. 'Bring me a wife, and your crops will thrive. Bring me nothing, and you will all starve!'

There is a sudden bang and a cloud of smoke where Owd Hob stands, and the smaller children scream and the adults laugh, as the boy scurries behind the scenery.

The backdrop is rolled away to be replaced by another one showing the inside of an old tavern. The farmer sits at a wooden table in front of it, his head in his hands.

Lucy, the narrator, says, 'All night, the farmer wrestled with his dilemma! He loved his daughter very much, but if the crops continued to fail, then it would be a disaster!'

Another girl walks onto the set in a black smock. She sidles up to the farmer and dips a hand into his pocket, drawing out a big gold-coloured coin. The farmer turns around and grabs her wrist.

Lucy says, 'But then the farmer has an idea. Owd Hob said he wanted a wife, not necessarily the farmer's daughter. And in this inn at a village some miles away, there was a girl who was known to be a thief, a liar and a cheat.'

'Nobody will miss you!' shouts the farmer. 'I'll be doing this village a favour! You're coming with me!'

The scenery changes again, to the hills and moors at sunset, the sinking sun painting the sky vivid reds and oranges. The farmer stands with the girl, who is bound by ropes and gagged with a rag, in front of Owd Hob.

'I have brought you a wife as you asked, Owd Hob!'

'She will do very well, Farmer!'

'And you promise our crops will thrive this year?'

'They will!' says Owd Hob, taking the struggling girl by the arm.

'And next year they will thrive also?'

'They will!' shouts Owd Hob. He turns to the crowd and winks. 'Provided I get another wife!'

The audience laughs, though Sophie is not quite sure what she is meant to find funny.

Owd Hob and the girl disappear in another puff of smoke. Lucy says, 'And that is why Withered Hill is blessed with a good harvest every single year! The end!'

The audience applauds and Owd Hob, the girl, the farmer and his daughter all come to the front and hold hands in a chain with Lucy at the end, and bow down.

'Did you like that, Sophie, love?' says Carol, leaning across Noah as she claps loudly. 'It's a lovely story, isn't it?'

'I suppose,' says Sophie, uncertainly.

'Well, it's your turn now!' prompts Carol. 'Guest of honour this Beltane!'

The crowd falls silent again as the scenery is wheeled away, revealing the big pile of grain sacks. The members of the parish council all stand and Carol takes Sophie's hand, and they lead her in a procession towards it.

'What am I supposed to do?' hisses Sophie. 'Nobody has told me anything about this.'

'It'll all become clear, love,' whispers Carol, leading her to the pile of grain and seeds.

Sophie frowns. Is something moving in one of the bags, just about at head height, nestled in with all the others?

'Catherine's brought you something from the outside,' says Noah.

'Take a look,' says Carol, nudging Sophie.

Acutely aware of the silent crowd behind her, waiting expectantly, Sophie reaches forward hesitantly for the bag that is moving more quickly now. What is it? An animal? There's a muffled noise coming from inside, as though whatever is in there has just woken up. Sophie takes hold of the top of the bag and glances at Carol, who smiles and nods. She takes a deep breath and drags up the grain bag.

Sophie takes a step back. There's a man there, buried and hidden beneath and between all the grain sacks, only his exposed head now showing. And she knows him. She knows who it is, this man with bloodied cuts all over his face, one eye swollen and closed, his nose bent and broken, his mouth covered with a dirty old rag.

'Jamie,' she says.

14

Inside

Days in Withered Hill: 319

Over the months since Sophie had arrived in Withered Hill, Catherine has provided her with huge lever-arch files of printed pages from social media, which she devours, imprinting the memories onto her brain. @SophieWickham2. She wonders if there is a @SophieWickham1 out there. She supposes there must be, hence the username. How odd to have someone going around with your name. Sophie spends hours looking at the printouts, studying the pictures that had been posted. It is very odd to look at your own face and have no memory of taking the photograph. But, little by little, Sophie builds up those memories like a house, brick by brick, until when she flips open the file at a random page and sees a photograph of a dark-haired woman laughing with two other girls in a crowded bar, she can almost put herself there, taste the rum and Coke, hear the music playing, feel the jostle of bodies. Oh, she aches so much to be in a bar, surrounded by people, the night shivering with possibilities.

The chronological thread of Sophie's posts detail a gradual uncoupling from the wide circle of friends she had around her at the beginning of the social media record.

Jill and Pete are sooooo loved up! I think it's going to be time to buy a hat soon, what do you think, @jillyjoolie?

Huge congrats to my bestie @LizRog3rs, she's only having a baby!

Great to see all the girls again at @DonnaDonnaB's utterly fabulous going away party. Enjoy your new life in Dubai, babes xxx

By degrees, Sophie's friends are married, or have babies, or go away, and Sophie's world shrinks and shrinks and shrinks. And then Jamie comes along.

Reading their initial interactions on social media reveals the halting, cautious birthing of a relationship, their posts and messages bristling with sexual tension. Catherine has access to the private messages as well, and they are sexy and passionate and make Sophie squirm with longing as she reads them. But when Jamie and Sophie get together, there is a cooling of the online ardour – presumably transferred to real life – and then there are increasingly tense and snippy exchanges in the messages, and eventually just the public post: *Looks like I'm single again :(*

There are no more private messages or interactions. After the last exchange is the note: *You can no longer receive or send messages to @JamieH79 as this user is now blocked*

There is a black hole in the memory palace that Sophie has built from the hundreds of online posts. A black hole into which Jamie disappeared and never emerged from.

Until now.

'Sophie,' croaks Jamie, his lip split, his tongue swollen, as Catherine pulls the rag away from his mouth. He looks at her uncomprehendingly through his good eye. 'Sophie.'

Sophie stares at him, then looks uncomprehendingly at Catherine.

Jamie tries to look around but can't move his head very much, trapped in all the sacks and bags. 'W-where am I? What are you doing here?'

'You're in Withered Hill,' says Sophie. Dusk is falling rapidly, and the villagers are sitting quietly on their hay bales, or standing silently behind. 'I live here.'

Jamie doesn't understand. He looks like he might cry or fall asleep.

Sophie turns to look at Catherine. 'Why is he here? What happened to him?'

'I did,' says Catherine. 'He didn't want to come with me at first. So I had to use persuasion.'

'S-she attacked me.' Jamie gasps. 'She hit me.'

'Bit of a pussy, isn't he?' Catherine said, smiling.

Someone lights the bonfire and it bursts into yellow flame with a whoosh of burning accelerant. The light paints the faces of the villagers watching them.

Sophie says, 'Why did you bring him to Withered Hill?'

'To further your education.' Catherine turns to Jamie, who is starting to struggle inside the pile of sacks. 'Don't bother. You're tied up nice and good. Now, tell her.'

'Sophie? Tell her what?' moans Jamie.

'Tell her about you, and Sophie, and everything.'

Jamie cranes his neck, seemingly seeing the gathered villagers for the first time and the bonfire that is burning merrily now. 'What's happening here?'

Sophie flinches as Catherine slaps Jamie hard across his battered face. 'Stop asking questions, and start doing as you're told.'

'I'm sorry,' says Jamie, and bursts into hacking, snotty sobs. 'Sophie, I'm sorry. If this is some kind of punishment for how I treated you, I'm sorry.'

Catherine hits him again and he moans. 'Stop saying sorry. Tell her what you did.'

Jamie fixes his one open, teary eye on Sophie. 'All I did was love you.'

The bonfire crackles in the silence that Sophie weighs between them. She says, 'Tell me why.'

Jamie licks his cut lips. He asks in a dry whisper, 'Can I have a drink?'

Sophie looks to Catherine, who shrugs and puts a bottle of water to his lips.

When he's taken a long draught, he says, 'I loved you because you were perfect. Beautiful. Kind. Funny. Clever. And you were mine and I couldn't believe that someone like you would look at me twice.'

He doesn't look like much, it is true, his face all battered and bleeding. Sophie doesn't know where Catherine has taken him from, but she has been gone from Withered Hill for a couple of days. London, she imagines, approaching Jamie on the street and saying she needed to talk to him urgently. Then going back to his flat, or perhaps an alley in the night, and setting about him with a baseball bat or a house brick, kicking and punching him even as he curled in a foetal ball and begged her to stop.

'So why did you do what you did?'

Jamie looks at her for a long moment, trying to open his swollen eye. 'I didn't want to lose you.'

All the things she has read on social media have taken root in her mind. They have become memories, scattered like dandelion clocks on the spring breeze, or the sparks floating from the crackling bonfire. Jamie being in front of her, here, now, is the glue that holds them all together, as if someone is joining the dots between all his words and actions and creating a picture of how he really was.

'Another person isn't yours to lose, or keep,' she says. 'They're not a thing, a possession, a prize.'

'We were meant to be together,' says Jamie, almost too quiet to hear.

'Still, you moved on, didn't you, after you'd stopped harassing Sophie?' goads Catherine. 'Gave up eventually. Got yourself another girlfriend. What's her name?'

Jamie mumbles something.

Catherine splashes the water from the bottle into his face. 'Speak up.'

'Kara.'

'Kara,' says Catherine. 'Kara went out with friends last night, didn't she? And what did you do?'

'I was just worried about her. London's not a safe place for women.'

'Or men,' says Catherine mildly, raising an eyebrow. 'I'll ask you again. What did you do?'

'I followed her,' whispers Jamie.

Catherine nods. She looks at Sophie. 'And I followed him. Lurking in doorways, trailing her and her friends from bar to bar.' She shakes her head. 'Pathetic, controlling, useless, little man.' She squirts the last of the water in his face and throws the bottle at him. He winces as it bounces off his forehead. 'I've no idea what Sophie ever saw in you. You didn't deserve one inch of her.'

Sophie is aware that behind her the children have gathered, the younger ones who danced the Maypole and the older ones who performed the play. They are arranged in rows, and have started to hum, in tune, together. She says, 'What happens now?'

Catherine is passed something by a figure behind her. Mr Purcell, the butcher. It is long and flat and wrapped in black cloth. She hands it to Sophie, who unwraps it carefully. It is a long-bladed knife with a worn, leather handle. Evidently old, but keen; the blade glints in the orange flames of the bonfire.

'It is Beltane,' says Catherine simply. 'We must pay tribute.'

Mr Purcell is joined by Peter O'Keeffe, and they start to pull the sacks of grain away from Jamie, exposing his body. He is tied with lengths of rope around his wrists and ankles to some sort of wooden frame. As Sophie watches, the butcher and Peter grab each side of the frame and suddenly upend it, placing it so that Jamie is now hanging upside down, his head near the sacks of grain and seed scattered beneath him. Jamie looks around wildly, mewling like an animal.

Sophie looks at the reflection of her own eyes in the flat blade, then at Jamie. He sees the knife properly for the first time, and starts to moan and wail.

'You want me to...?'

Catherine nods. 'Spilling his blood on the grain will help us give thanks for what has been, and for what is to come.'

Sophie at last thinks she understands the play from earlier. The children behind her have started to sing a low, mournful tune, rising in pitch and gathering speed. She can't make out the words. When the character of Owd

Hob demanded a wife, he meant a sacrifice, didn't he? To ensure the harvest. An unworthy soul who no one would miss. She looks at upside-down Jamie, whose one good eye is wide, and he's shaking his head violently. He pulls at his bonds, but his limbs are stuck fast.

'Like a pig,' says Mr Purcell matter-of-factly. The children are singing louder. 'In at the sternum, then push up to sever the arteries. It'll be quick if you do it right.'

Jamie finds his voice at last, and starts screaming 'help help help' over and over again. The children are singing more loudly. And now the adults are chanting behind Sophie. 'Kill the pig! Kill the pig! Kill the pig!'

Will anyone miss him? Perhaps Kara will wonder where he went, then quietly move on, with a sense of relief. Jamie has parents, she thinks she remembers; she's sure he mentioned them in his social media posts. Perhaps they'll miss him. She looks at the knife again. But isn't it all rather moot? Can she do this? Really? Does anyone deserve this?

'I'm sorry,' says Jamie hoarsely. 'I'm sorry I hit you.'

Sophie's eyes narrow. 'What did you say?'

'When you met that man. After chatting to him online.'

The voices of the children swell behind her. The adults are still calling for her to kill the pig. She glances away for a second, meeting Catherine's steady gaze. It is properly dark now, the flames from the bonfire licking the black sky. She still can't make out the words the children are singing, but they are filling her with joy, and hope, and the sense of both endings and beginnings.

'Say it again. Tell me.'

Jamie licks his lips, but no more water is offered to him. 'You met that man. When you came home… why are you making me say all this?'

'Tell me,' hisses Sophie.

'When you came home… I showed you your messages. I'd hacked into your account. You said you wouldn't delete your socials. So… so I hit you.' Tears are flowing down his bruised face. 'I'm sorry. I tried to tell you I was sorry. Dozens of times. I sent you flowers and presents and I tried to see you at your flat and at your office, but you wouldn't talk to me.' He starts to cry. 'Please let me go. I won't say anything about all this. I'm sorry I hit you.'

Sophie looks at the knife in her hand. She looks at Catherine, but her face remains impassive. Sophie turns and beholds the villagers, all looking expectantly at her.

'You are a pig, Jamie,' she says. The villagers are chanting louder and faster, drowning out the song of the children.

Sophie squats down and puts the point of the blade at Jamie's neck, just beneath his bobbing Adam's apple.

'That's it,' says Mr Purcell encouragingly.

'Kill the pig,' says Catherine.

'Straight in, then up,' says Peter O'Keeffe.

Sophie puts pressure on the knife, seeing the point press into Jamie's soft skin.

Then she takes it away.

'No,' she says. 'I can't do this. No matter what he's done.'

Catherine gives her a tight smile and takes the knife from her trembling hands.

'I'm sorry,' says Sophie. She takes a deep breath. 'There'll be no sacrifice. I'm sorry to you all.'

Catherine smiles at her, and crouches down, looking into Jamie's terrified eyes. 'There's always a sacrifice on Beltane,' she says.

Then, in one swift movement, Catherine forces the knife into Jamie's throat, and expertly twists it. A gout of blood splashes out, onto Catherine, onto the sacks of seeds and grain and Sophie's feet. Her eyes widen in horror, and as the villagers start to cheer, she runs, pushing through them, into the night.

15

Outside

Days to Withered Hill: 9

Not far from Sophie's flat, there was a church with a small tumbledown graveyard wrapping around three sides of it. She'd never actually seen the church open; it was probably out of use, its dwindling congregation subsumed by a larger competitor, but it seemed to be kept in a fair state of repair. The small cemetery, though, was often overgrown, save for some weeding and mowing twice a year, and it was here Sophie found herself, stepping carefully in her heels over the weeds and nettles encroaching onto the gravel path.

She had been coming for a few months, ever since she had been forced to stop seeing her therapist because she couldn't justify the cost any longer. She had an uncharitable thought which she waved away, but it was one that was coming more and more often these days, and getting more difficult to dismiss. *How long is she going to hang on?*

When her parents had died, it turned out that they didn't have as much money as they had liked to pretend they did to their friends. In fact, they were in quite substantial debt, and it seemed her father had had something of an online gambling habit. The creditors were lining up for their share of the life insurance payout, which Sophie had to dole out in the weeks after the crash. All that was left was the house, and while it would realise a fair amount in the current market, there was one rider that had been written into her parents' will.

It was to be put into the name of her gran and to be her home until such time as she could no longer live independently in it, or passed on.

All of which meant that Sophie was left with pennies after her mother and father died. She had subtly tried to suggest to her gran that she might be happier in a smaller flat, and they should maybe sell the house and buy one outright, but she was having none of it. Sophie didn't think she realised how broke they actually were, with the house their only asset. When she had finally decided to go back to university, she had to get student loans and three years later came out of it with debts up to her eyeballs, while her gran swanned around a house that was so much too big for one person that it was actually quite obscene.

How long is she going to hang on?

There was the thought again. Because her parents' will had also said that Gran wasn't to sell the house without Sophie's permission, and that after her death, the property would revert to her. She was constantly on house price websites, watching the value of it go up and up. When she got that house, she would be sorted. There'd be a for-sale board up before Gran was in the ground. She'd been looking at the prices yesterday, and then another thought had popped into her head. Not how long is she going to hang on, but what if… what if… what if…

That was why she'd had the dream last night. And why she'd decided to come here before she went to meet Tom.

When Sophie moved to London, her gran had given her a small chrome vacuum flask. Sophie had been touched, thinking it was tea for the train journey. It wasn't. It was a portion of her mother and father's ashes.

Sophie had been appalled. 'Not all their ashes,' said Gran. 'But just enough. So you can remember them properly. You won't be able to get up here much to see their marker in the churchyard when you move to London.'

For months and months, the flask had been hidden at the bottom of a drawer that Sophie never went into. But on the day that she'd told the therapist that she couldn't come to see her any more because she couldn't afford it, she'd dug it out and stared at it. Then she'd gone for a walk, ending up in the graveyard, and on a whim she'd emptied the dust out at the base of a tree. She'd taken with her four cans of pre-mixed gin and tonic and stood there for a long time, drinking silently, and thinking. She'd wondered what her actual parents would say, if they were still alive. Three cans of gin and tonic down, she'd started to talk to them. The pile of ashes, of course, didn't talk back. But then, neither did her therapist, other than to tell her the hour was over, pushing the card machine over the desk. Her parents' ashes listened, that was the main thing, and they didn't cost anything.

Now she stood in front of the same tree, the ashes long ago washed away or absorbed into the earth, imagining her mum saying to her *You look nice.*

'Thank you,' she said aloud. 'I have a date. But...' Sophie glanced around the deserted graveyard. 'Mum, I feel like I'm becoming *unmoored* again. Last night, I had a dream...'

-

'It wasn't a hare at all,' said Sophie in a whisper to the tree. 'It was Emily. Of course it was.'

She hadn't thought about it for years. For twenty years. For more. Ever since it happened, really. She had been sitting watching TV with Gran when Gran had decided she needed to go and put on her pyjamas and dressing gown. Sophie had sat on the big sofa and watched the baby monitor lights suddenly flash as Gran must have gone into the nursery. Then the lights had glowed so bright that Sophie had thought the tiny bulbs might explode. Gran was screaming.

When Gran had eventually come down, ashen-faced, and phoned Sophie's mum at the party, though she kept her voice low, Sophie had heard Gran tell her mum that she thought Emily was dead. Even from the sofa, she could hear Mum screaming down the phone, and Gran had put down the receiver and then called an ambulance. Sophie had cried and cried and cried and cried until it came.

Oh God. Why had that dream happened? It was that stupid hare that had arrived at work. And the thoughts she'd been having yesterday. About Gran. And what would happen if… what if… what if…

'The horrible thing was, Mum,' whispered Sophie, 'I wasn't that sad, really. Not after it had all sunk in. Emily was a baby and didn't feel like a real person. She just laid there and cried and slept and ate, most of the time.' Sophie remembered her mother crying for days, her father walking around the house as though he was never quite sure why he had just entered a room. 'I couldn't even understand why you were sad. You still had me.'

She stood there in silence for a while, taking deep breaths. She hadn't thought about Emily dying for ages.

Months. Every time she did, she felt odd. She sighed, long and raggedly. 'I'm becoming *unmoored* again. I can feel it.'

–

After university, Sophie had been determined to sort her life out. After spending six months at the house with Gran, she decided to move to London, where most of her friends were based.

Sophie discovered in London she had a talent: being able to forget everything bad that had happened to her. Emily, and even her parents' deaths… she was able to lock them all in a box in her head and leave them there, untouched and untroubling her. She got a job with an events management company on a one-year contract, and she helped to organise gigs and performances around London. When that ended, she signed up with a temping agency while she looked for something more permanent, but she seemed never to find the time to apply for anything. She was having too much of a good time, and so long as she was making enough to cover her rent, buy food and go out to the pub, she was happy enough to drift along.

Besides, all the gang was there from university. Jill, Donna and her very best friend, Liz. They were a fabulous foursome who spent their weekends doing shots and dancing and pulling unsuitable men. They scrimped and saved to go to Ibiza for their holidays. Sophie didn't care to find a career that she'd have to throw herself into and devote time and energy to – she was too busy doing that with her friends.

Jill was the first one. She took a boy called Pete home one Saturday night, and then arranged to meet him again.

Sophie was aghast. That was not what the Fabulous Four did. They met them, let them buy them drinks, took them home, shagged them, then threw them away. Jill had ignored the code. She had broken the rules. Still, thought Sophie, it'll never last. But it did. And slowly, Jill stopped going out as much with the rest of them, and started going out more with Pete. And suddenly six months had gone by, and Sophie was listening with a rictus smile plastered to her face while Jill gushed and showed off the engagement ring.

> Jill and Pete are sooooo loved up! I think it's going to be time to buy a hat soon, what do you think, @jillyjoolie?

Still, there were three of them, and three was a good number for a night out. And they did as they did before, drinking three, four nights a week and planning their next holiday, while Jill and Pete were planning their wedding. And life was still good.

Sophie even got to love Pete, and enjoyed their wedding immensely, even when Jill told her that they were moving out of London and going to buy a home near Pete's parents in Devon, and shouldn't she seriously think about getting out of the city once in a while? Sophie had stared at her in disbelief. Leave London? Why would anyone do that?

It was at Jill and Pete's wedding that Liz, her very best friend, got off with the best man. Which was a major bit of gossip that had Sophie cackling for hours the next day, telling her bestie how much of a cliché she was. But the laughter faded when Liz decided to see Andy again. And then again. And then, a year later...

> Huge congrats to my bestie @LizRog3rs, she's only
> having a baby!

And then there were two. Until the day of the christening of Liz and Andy's baby, Harry, at a little church out in the country that took an age for Sophie and Donna to get to. It was summer, but it was pissing down. Sophie got hammered in the pub after the ceremony, leaning on Donna as she had a cigarette under a leaky verandah at the back of the pub, looking out at the torrential downpour.

'Remind me why we live in this country again, Donna?' she slurred.

'I was going to talk to you about that,' said Donna carefully.

> Great to see all the girls again at @DonnaDonnaB's
> utterly fabulous going away party. Enjoy your new life
> in Dubai, babes xxx

And then there were none. Without the reflected light of the Fabulous Four, Sophie's life suddenly looked lonely and sad. She was drifting, watching from a distance as her friends made lives for themselves.

–

Sophie stared at the tangled roots of the tree, emerging from her reverie. She had talked about the dream, about the memories of her dead sister. Now she felt cleansed. She locked Emily away back in the box in her head, and almost immediately forgot about her, about what had happened.

That was her talent. Sophie Wickham lived in the moment, in the now; right at this moment, she had a date with Tom Gisburn. She thanked her parents for their time, smoothed down her clothes, and turned to leave.

This time, she thought. *This time, it's finally all going to go my way.*

16

Inside

Days in Withered Hill: 319

Sophie runs for her cottage, but doesn't make it, falling to her knees on the grass in front and throwing up repeatedly until her stomach is empty.

Catherine just killed a man.

Catherine just killed Jamie.

Her stomach twists in knots and she heaves, but nothing is left. She retches so hard, her head spins and she sees stars dancing in front of her eyes, and pitches forward to lie in the cool grass, her hair in the pool of vomit.

'Hey.'

Sophie doesn't lift her head. 'Cat.'

She feels Catherine crouch down beside her and lay a hand on her shoulder. 'We should talk.'

Sophie opens her eyes and stares at the grass, every blade in sharp focus in the clear night. 'About the fact you just butchered my ex like a pig?'

'Yeah, about that. We going to do it here or shall we go inside?'

Ten minutes later, they are sitting on Sophie's couch, nursing large glasses of whisky. Sophie watches the ice cubes melting and says nothing for a long time. Then she looks up at Catherine, her face painted silver by the Beltane moonlight flooding in through the open curtains, and says, 'You're a murderer.'

'I killed a man, yes,' says Catherine. She takes a sip of her drink. 'Murder... that's a concept. An idea. A crime. It is something enshrined in the law of the outside. And the law of the outside is not quite the same as the law in Withered Hill.' She looks at Sophie for a long moment. 'So, why couldn't you do it? Why couldn't you cut Jamie's throat?'

'Because it's wrong!' exclaims Sophie. 'It's *murder*... I can't believe you even asked me to.'

'A question,' says Catherine. 'What's the difference between killing Jamie, which you refused to do, and burying Zachary alive at Faunalia, which you participated in?'

Sophie thinks about it. 'It was what Zachary wanted. He would have done it anyway, slowly and with a lot of effort.' She looks at Catherine. 'It quite obviously wasn't what Jamie wanted.'

'Did you not even consider doing it? Not even for one moment?'

Sophie thinks about how she was feeling with the knife in her hand. Triumphant. Vengeful. Exhilarated. And something else, as well. As the crowd pressed forward, as she felt their collective presence at her back.

'For a moment,' she admits, 'I felt... part of something.' She drains her glass. 'I felt connected.'

'And not just to Withered Hill,' says Catherine, pouring Sophie another whisky. 'To everything.' She sits back and watches Sophie for a while. 'Nothing really dies. Not really. Everything is linked. Connected. There is a... force that moves through all of us. It is the wind through the trees and the heat of a sun-warmed stone and the beating of our hearts. It is within us and everywhere around us. And that is the essence of life in Withered Hill. That nothing ends, that there is a continuous cycle of life and death and rebirth. It used to be the way of things the world over, but gradually people forgot this. Out there, they live in the moment, forgetting the past and ignoring the future. They have forgotten that they are part of a much broader canvas, that spools out in front of them and behind them into infinity.'

It is the longest and most serious speech Sophie has ever heard from Catherine. But it does not go anywhere close to answering her questions. 'But you killed Jamie. That was wrong. How does that not make you as bad as them out there? Worse?'

Catherine refills their glasses again. 'Because it is about balance.' She puts her glass on the arm of the chair and holds her hands out, palms up. 'In here, I hold Jamie's life. A life badly lived, pockmarked with lies and violence. In here, I hold the good fortune of Withered Hill, dependent on the harvest.' She weighs her imaginary ballast in her hands. 'The needs of the many, balanced against the crimes of one.'

'But who are we to judge?'

'We don't judge. We follow the natural law. And that says a sacrifice of the unworthy must be made at Beltane to ensure the harvest.' Catherine shrugs. 'We are all just instruments through which the life force flows.'

Sophie stares out of the window at the black night. 'Then... I failed. You asked me to do it, and I couldn't. What does that mean?'

Catherine smiles. 'No, you didn't fail. You passed. With flying colours.'

Sophie gapes at her. 'It was a *test*?'

'Of a sort. Had you killed him, you would not have been ready to depart Withered Hill. To go outside. Not yet. Perhaps not ever.'

'And yet, Jamie's still dead.'

Catherine moves over to sit with her on the sofa. 'I had to maintain the balance. I took something bad and unworthy and utilised it to make things better for the majority.' She puts her hand on Sophie's knee. 'Don't think of it as killing a person. Think of it as redistributing the force that winds through us all. Using it for the greater good, rather than having it stoppered up in a sad, vicious little man.'

Sophie thinks about the way the children were singing and the villagers were chanting. About the look in Catherine's eyes, the shining light there as she thrust the knife into Jamie. 'You... all of you... you're different at your festivals. You frighten me.'

'Still?' says Catherine. 'After almost a year?'

'Still,' agrees Sophie. 'And the thing that frightens me most is... I don't know whether I'm becoming like you, or will never be like you.'

Catherine puts a finger under Sophie's chin and lifts her face up and towards her. She pushes a strand of hair over her ear. Then she leans forward and kisses her.

'Ugh. You taste of sick,' Catherine groans.

Sophie bites her lip. 'Sorry. Should I clean my teeth?'

Catherine says nothing, just takes her by the hand, stands up, and leads her towards the stairs.

Days in Withered Hill: 323

Two weeks after Beltane, Sophie takes a walk in the warm May sunshine down to the farms crouched on the inside of the woods. Up on Peter O'Keeffe's top field, there stands a new scarecrow. Sophie has avoided it for the past fortnight, but now she feels ready.

She waves at Peter's wife, Jan, in the yard with the pigs, and lets herself into the field, closing the gate behind her. The ground is dry, and she walks along the parched tractor tracks towards the scarecrow, crucified on a wooden frame amid the swaying wheat. As she gets closer, she sees it wears a pair of shapeless brown trousers and a checked shirt, its face a rough hessian sack, topped with a floppy sun hat.

Flies buzz about the scarecrow, and Sophie wrinkles her nose and wafts them away as she stands before it, the sun behind casting its shadow across her. She reaches up and takes off its hat, using it to bat at the cloud of flies, then stands on her tiptoes and pulls off the sack.

Jamie's sightless eyes stare back at her, his face bloated and mottled with blue and black patches. Something like fungus seems to grow at the corners of his mouth and his nose. His eyes are white and glassy, sunken in his face. He stinks. Flies play about the gaping wound of his slashed throat. Something moves and writhes in there. Maggots.

'Did you deserve this, Jamie?' she says aloud. He doesn't answer, of course.

Even after almost a year in Withered Hill, Sophie doesn't think she'll ever fully understand it. She keeps being told that she is close to cracking the puzzle that will allow her to leave, but she has no idea if she even wants to go out there any more. What would have happened to Jamie on the outside? For hitting Sophie? A police investigation. A court case. Perhaps a small prison sentence. Or perhaps he would have used his coercion and viper tongue to convince a jury that he was innocent, that if it did happen, then, somehow, Sophie deserved it. Maybe he would have become a reformed character. Maybe he would go on to do it again, and again, and again, but worse and more violent each time.

Who knows? Not Sophie. All she knows is that Jamie's body was tied to a wooden frame and planted in the field, and that his body is slowly rotting and decomposing, and his flesh and skin and bones will slide off the frame and be absorbed into the soil and nourish the earth until there is nothing left of him at all.

'I'm not sure you did deserve this,' she muses. 'But neither did you have any right to hit anyone, or try to control them. Perhaps if you knew what awaited you, what your punishment would be, you wouldn't have done it.'

Flies buzz around Jamie's mouth. Sophie swats at them with the hat. She sighs. 'That's the way of things in Withered Hill, as people keep telling me. When I'm out there...'

When she's out there, outside, in the world, she won't be in Withered Hill, won't be subject to its natural law. But perhaps, having seen it, having seen what can happen to people... maybe she'll be *better*. Maybe that's what all this is about. Seeing the bigger picture. Knowing that actions have consequences. Realising that those consequences might not be the ones you were expecting.

'It's too late for you, Jamie,' she says. 'But not for me.'

Then she reaches up to pull the hessian mask over his face, and replaces his hat, and walks away from him, back down the field.

At the gate, Catherine is waiting for her, in a light sundress. She greets Sophie with a kiss. 'Made your peace?'

'Sort of,' says Sophie. She doesn't look back at the scarecrow. 'I think I understand. A little, at any rate.'

'Good,' says Catherine. 'It's a lovely day. Treat you to lunch at the pub?'

Sophie nods, and allows Catherine to take her hand.

17

Outside

Days to Withered Hill: 9

The Green Man had a single line of tables outside on the pavement behind a barrier, under the trailing plants that grew in long boxes above the windows, ivy wrapping around the pub sign displaying a sylvan study of a man's face formed from leaves and branches, berries and nuts. The pub was just off the main road and the warm evening meant that every table was occupied. And then, at the far left, on the last table, a figure stood up and Sophie realised it was Tom Gisburn.

He was wearing light chinos and a fitted summer jacket, over a black shirt open at the collar. As she walked over, she saw his eyes were just as piercing a blue as she remembered from their brief first meeting. His beard was closely trimmed and his hair had a salting of grey. He smelled woody and earthy.

He stepped outside of the barrier to greet her, putting his hands on her elbows and giving her a chaste kiss on the cheek. She felt herself twitch, nonetheless, and inhaled his scent. He took a step back and admired her, and said, 'I'm glad you got in touch. I really do need to recompense you for my clumsiness.'

Tom pulled a chair out from the round table and indicated for her to sit down. He looked up and raised a finger, and a waiter practically ran over, as though compelled by his magnetism.

'I really don't want any money,' said Sophie as Tom sat down. She smoothed her dress self-consciously; it felt cheap next to Tom's understatedly expensive designer labels. 'They washed OK.'

'Then tonight is on me,' he said as he sat down. 'I'm going to start with a beer. What do you want? Have you eaten? They do an amazing charcuterie board if you're not so hungry, or we can get the full evening menu if you're peckish.'

Suddenly she'd forgotten what she liked to drink. 'Gin and tonic,' she told the waiter. She didn't even like gin. She wanted a glass of wine. What was wrong with her? 'The charcuterie board sounds great,' said Sophie as the waiter scurried away.

Tom's brow furrowed. 'Hmm. I just realised that made me look a bit cheap, suggesting the sharing platter first. Shall we get the menu?'

'Honestly, it's fine,' she said as the drinks arrived. 'I'm not that hungry.'

Tom ordered and then turned back to Sophie, fixing her with his eyes. 'What do you do in that office I steamrollered you in front of?'

'Data management,' said Sophie. 'It's exactly as dull as you would imagine.' She thought of all the zeroes and ones. And the hare. 'Well, mostly.' She sipped her drink. 'I also volunteer at a charity shop near where I live. Just a few hours, every couple of weeks.'

'How noble,' said Tom, smiling.

Sophie shrugged. 'It's no big deal. I just like to do my bit, you know.'

Tom's eyes were on her, and Sophie felt a little uncomfortable with his scrutiny, though she couldn't really pinpoint why. She said, 'What about you?'

Tom shrugged. His shoulders were nice and broad under that jacket. 'Consultancy, mostly. Bit of this and that.'

Sophie took a sip of her drink. 'Very mysterious. I suppose that means you're actually a spy, then.'

Tom laughed, a rich, soulful sound. 'I'm really not trying to be mysterious. What I do is probably more boring than what you do. So let's not talk about work at all.' He leaned back in his chair, as though sizing Sophie up. But not in a predatory, creepy way. Not like Colin looked at her in the office. Just like someone who was... interested. 'Tell me about you. Where are you from?'

'Small village in Gloucestershire. You won't have heard of it and you'd forget the name straight away if I told you.'

'Parents still there?'

Sophie took another drink. 'No, they're both dead.'

He bit his lip and frowned. Sophie found it quite exercising. 'God, I'm sorry. That must have been tough on you.'

'It was,' she said.

'Any brothers or sisters?'

Sophie looked down at her drink. 'No,' she replied. 'No brothers or sisters. You?'

'I come from a huge family,' said Tom, grinning. 'I lose track of them, sometimes. No idea who is where or doing what.'

'Are you all spies, then?'

Tom laughed. It felt good to hear someone laugh at what she'd said, not because they were being sarcastic or insincerely indulgent, but because they genuinely found it humorous. Jamie had always put her down if she tried to make a joke, rolling his eyes and muttering 'You're not funny, you know'.

The waiter arrived with the platter, which did look delicious as he set it down between them. 'Another round of drinks, I think…?' said Tom.

After they'd eaten, they moved inside the pub, at Tom's insistence, as the temperature was falling along with the night. He bought more drinks but was careful to keep asking her if she'd like a soft drink or water, making great pains to show that just because he was footing the bill he wasn't trying to get her drunk. Sophie's own promise to herself that she wouldn't drink too much had gone out of the window after the third gin. But she was bubbly and happy and the conversation flowed, sometimes so fast that she couldn't even keep up with what they were talking about, or remember it five minutes later. She kept saying funny things, and Tom kept laughing, and at no point did she find him staring at her tits or accidentally putting his hand on her thigh.

At eleven o'clock, he looked at his Rolex and glanced at her. 'I think that's us for the night. Let me get you an Uber.'

'No, honestly, I'm literally round the corner.' Sophie felt suddenly crestfallen that the evening was over. This was the bit where he told her that he had to get home before his wife got in, or that he was up early to see to the children. He caught her looking at him and laughed.

'In case you were wondering why, it's because I've got a six o'clock video call with Japan.'

'I believe you, Mr Bond,' said Sophie in a stupid accent that made him chuckle anyway.

He stood up and took her elbow. 'Come on, I'll walk you to your place. I'll order a car on the way.'

They chatted as they walked, their arms touching. Sophie had made another promise to herself before she went out, too. She would absolutely, definitely, in no uncertain terms, *not* sleep with Tom Gisburn on their first date.

They stopped outside her building. 'It's been lovely,' he said. 'I don't suppose you'd want to do it again…?'

Sophie leaned back against the wall, and bit her lip, and looked up at him. She took hold of his hips and pulled him into her.

'Sophie…' he said gently.

'Shut up and kiss me,' she said.

Inside the flat, Sophie stood on the rug in the centre of the living room and carefully unbuttoned the front of her summer dress. Tom stood by the door, watching her.

'Come here,' she said.

He took a step forward, and another, not taking his eyes off her. Then he paused at the edge of the rug and glanced down at his feet, then back at her. 'You come here,' he said, raising an eyebrow.

Sophie bit her lip again. She liked it when he used that tone of voice. She smiled and walked towards him in as sultry a manner as she could manage.

Tom Gisburn had never, Sophie hazarded, had to read a book called *How To Be A Considerate Lover*, nor did he run through a mental checklist of what he imagined her pleasure to require. He just knew instinctively. There was something animalistic yet controlled, gentle yet dangerous, abandoned yet measured, about his lovemaking. Sophie surrendered to him completely, and let him consume her, and she fell asleep when they had finished, sated and satisfied, in a tangle of limbs, inhaling his woody scent.

—

Sophie had set her alarm early so she could shower before work, and see Tom off. When she woke from a dreamless sleep, the bed was empty, and for a moment she wondered if she'd just imagined it all. But his cologne clung to the sheets, and she wrapped one around her and slid out of bed, grabbing her phone and walking to the kitchen to make a coffee.

He had gone but had left one of his business cards on the kitchen worktop, with the words CALL ME, followed by a single X. Smiling, she made a coffee and headed into the living room, and screamed.

The cockroach scuttled under the big rug. God, she hoped Tom hadn't seen one or he'd think she lived in a pest-infested hovel. Putting her coffee down, she grabbed her shoe and cautiously lifted up the corner of the rug.

All thoughts of insects fled her mind. Sophie frowned, and lifted the rug higher, then dropped it and shouldered the coffee table off it. She pulled the rug back totally, exposing what was underneath.

A five-pointed star, six feet across, carved into the wooden, polished floorboards. There was a double-edged circle surrounding it, and between the borders were symbols she could not understand.

Sophie tried to think back to when she'd last lifted the rug. Had she ever, since she'd moved in? Could this thing have been here all this time? And if not... what did that mean?

She jumped and shrieked as her phone buzzed in her hand. A direct message. It was from @coldiron6239745. He'd contacted her again, managed to message her even though she'd blocked him and hadn't added him to her friends list. And she knew that when she clicked on his name, his account would have been deactivated again.

She opened up the message. It said simply: THEY HAVE FOUND YOU.

18

Outside

Days to Withered Hill: 8

After Emily died, a policewoman spoke to Sophie alone for a long time in the living room. She didn't wear a uniform, which Sophie thought a bit odd at the time, and she kept looking at her mother, unsure if she should be with the strange lady on her own. But her mother was inconsolable, her eyes red-rimmed, her nose constantly running, her mouth a jagged red slash that just made form-less sounds of anguish, so eventually her father nodded abruptly and Sophie went into the room with the police lady.

She asked Sophie lots of questions, which she did not quite understand the relevance of in relation to what had happened. The lady asked Sophie if she was happy at home, if she enjoyed school, if she had lots of friends, if she liked being with her mum and dad at home, what Christmases and birthdays were like. She asked Emily if her gran was ever forgetful, or ever got angry with her. She wanted to know if Gran had been upstairs in Emily's room for any length of time.

Sophie told the lady that she had not looked in on Emily since Mother and Father had left the house, and she wasn't sure if Gran had. She pointed at the baby monitor and said Gran would have heard anything on that if there had been a problem. She told the lady she had sat with Gran on the sofa watching television. Gran had fallen asleep, and Sophie had woken her up to ask if she could have a snack, and Gran had looked at the time on her tiny wristwatch and then said yes, she would get Sophie some biscuits and a glass of milk, but first she had to check on the baby. Then Gran came downstairs looking flustered but putting on a fake, beaming smile, and asking Sophie if she could help her dial Mother's mobile number, and when the ringtone sounded, she went into the kitchen and shut the door. Sophie never got her biscuits and milk.

The police lady nodded. Sophie guessed she'd already got that story from Gran, who would have tried to under-exaggerate the amount of time she'd dozed on the couch and would have said that Sophie was with her all the time while they watched television. Instead, the police lady said, 'Sophie, did it bother you when Emily came along?'

Sophie made a show of frowning. 'I wanted a little sister more than anything!' she said, giving the lady her best winning smile.

'You weren't worried that mummy and daddy might not... have as much time or love for you, when the baby came along?'

Sophie pulled a face. 'Mummy said that she had enough love for all of us.' Which was true, she had said that when her belly was big and hard, before Emily came along.

'But now it's just you and mummy and daddy, isn't it?' said the lady softly.

Sophie nodded sadly. She thought about when Bambi's mummy was killed by the hunter, and felt tears prick her eyes. She let one roll down her cheek before wiping it away. 'I'm sad,' she whispered. 'I want Emily back.'

The police lady gazed at Sophie for a long time, with a look that the girl couldn't read. Then she folded up her little notebook and fastened it with an elastic band attached to the back. 'Very good, Sophie,' she said. 'I'll let you go back to your mummy and daddy now.'

'Thank you,' said Sophie politely.

The lady went to the door and paused before opening it. She turned around, as though she'd forgotten something. She looked at Sophie for quite a long time, long enough to make her feel uncomfortable. Then she said in a low voice, almost a whisper, 'Sophie? Do you know what happens to bad girls?'

Sophie wrinkled her brow to show she was thinking about it. 'They go to prison?'

The police lady nodded. 'Yes. Sometimes. Sometimes they go to prison. Sometimes, very bad girls who continue to be bad, though… well, something else happens to them.'

Sophie put her head on one side pleasantly. 'Oh? What?'

The police lady bared her teeth in what might have been a smile, but was very unsettling. 'Sometimes, Owd Hob takes them for his wife.'

'I'm sure I would not want to be married to anyone called Owd Hob,' said Sophie uncertainly.

'Those that do are generally better people on the other side of it,' said the police lady. Then she smiled more normally. 'Goodbye, Sophie.'

A few days later, Sophie's mother and father had a visit from a doctor, who gave them a piece of paper. It said that Emily had likely died from something called sudden infant death syndrome. Sophie wasn't quite sure she understood what it meant. She had not understood what the police lady had meant by what she had said before she'd left either, but the words made her uneasy and a little frightened.

The night Mother and Father were visited by the doctor, Sophie dreamed about Emily, except Emily was a hare wrapped in blankets, which made what was to come somehow a little less disturbing. It was only a silly old hare. Not a baby at all. When she awoke the next morning, she felt cleansed, though she couldn't have put that into words. She had forgotten what the police lady had said about what happened to bad girls, and she wouldn't have the dream about the hare again. At least not for more than twenty years.

–

It was all Sophie could do to concentrate on her work. At least it was Friday. And, she thought ruefully, she'd have the whole weekend to fret about that star carved in the floorboards under the rug. She tried to focus on the data she was entering, a series of telephone numbers of, as far as she could ascertain, particle physicists in several countries across Europe. But that direct message… THEY HAVE FOUND YOU. Who had found her? And why were they looking for her? And what did they want from her? And was the person who sent the message on her side, warning her? Or gloating?

Sophie pushed all the thoughts away and concentrated on the list of telephone numbers. She wondered if a particle physicist would have the answers to her questions, would be able to apply cold, objective science to her problem, see what she could not?

She worked until lunchtime, then picked up her bag and hurried outside into the sunshine. Colin glanced up as she scurried away from her desk, and seemed to be about to say something, but she ignored him. He would want to know how her date went, would be angling in his clumsy way to find out if she'd slept with her mystery man. As she emerged from the office onto the pavement, she had a sudden flash of Tom twisting her hands behind her back, holding her wrists in one strong hand, turning her over onto her front. So different from…

Sophie felt the colour drain from her face. Jamie. So different from Jamie. Was it him? The star carving? Had he somehow got access to her flat? Had he been in… when? While she was out with Tom? While she was at work? She racked her brains to try to remember the last time she had lifted that rug. She couldn't. The carving could have been there for days. Weeks. Months, even. Sophie felt her mouth actually fall open as realisation dawned. Jamie could have done it that very night, when she met @CallMeSir, when he hit her.

She sat down on a bench and took out her phone. She called up Jamie's profile and bit her lip as her thumb hovered over unblock. It wouldn't hurt, just for a quick look. Not that she knew what she was looking for. Having decided, she unblocked Jamie, and his feed unfurled before her eyes.

She wasn't quite sure how to feel at first, all those photos of Jamie with… Sophie clicked on the tags. Kara. Good-looking. Blonde. Petite. Pretty much the opposite of Sophie, looks-wise. There were pictures of them in the Lake District and on a Cornish beach and in front of the Arc de Triomphe and sipping extravagant cocktails beneath a palm tree. She felt a pang of something she didn't recognise.

He's moved on, she told herself. *He's got a new girlfriend. Kara.* Sophie exhaled, and felt as though the wind had suddenly been taken out of her sails. Then she laughed, abruptly and unexpectedly. Kara. Poor fucking her. It wasn't jealousy or regret Sophie was feeling. It was pure, unalloyed relief. Jamie had moved on.

She laughed again, and an old man walking by, leaning heavily on a cane, frowned at her.

She should warn this Kara, actually. Tell her what Jamie was like. She clicked on Kara's profile. It was set to private. She switched back to Jamie's. Well, if he was in the midst of an obsession with another woman, surely he wouldn't be breaking into her flat to make strange carvings in the floorboards. If he'd done it months ago, then he'd not followed it up with any more weirdness. She scrolled down his feed. He hadn't posted anything for quite a while, but that wasn't unusual with Jamie. He didn't use social media like normal people, for posting pictures of their dinner or their pets. He used it to keep tabs on his girlfriends. Sophie bet that if she could see Kara's feed it would be full of posts replied to by Jamie, staking his insecure claim on her every chance he got.

Sophie went to get a sandwich and coffee, then at the last minute swerved into the vegan place and got a salad and green tea. She was paying for the gym every month and hadn't been since... she blanched at the thought. Maybe she'd go this weekend. It would stop her thinking about the carving and the messages from Cold Iron. If the carving was Jamie, it was something stupid he'd done ages ago, not recently. Which, of course, didn't explain the perfectly timed THEY HAVE FOUND YOU. She felt so desperate to push it all away that she was even prepared to accept that the message dropping in when it did was just a horrible coincidence.

The rest of the afternoon passed quickly, and it was only when Colin was climbing into his grey, shapeless coat, that made Sophie itch and sweat just thinking about wearing the thing on such a beautiful, sunny day, that he dragged his feet over to her desk as she was packing up.

'How did your date go?' he said, brushing his fringe out of his eye.

Sophie continued to put her things in her bag without looking up at him. 'Do you want all the gory details, Colin?'

He paused and she glanced at him, licking his dry lips. He probably did, actually. The thought turned her stomach. She wondered if he'd ever had sex. That brought with it an unwelcome image of Colin naked, looming over her. The green tea made a slight, hot return to the back of her throat.

'Doing anything nice at the weekend?' he stammered instead of replying to her question.

Trying to discover who's carved a five-pointed star in my floor and finding out what it means, she thought, then paused. Colin had seemed to know stuff about the corn dolly, hadn't he? He was a bit geeky. Was he into all that… folklore stuff, too? She half-considered telling him about the star, and then her phone pinged on the desk.

She read the text message then gave Colin a glowing smile. 'Actually, yes, I am. It looks like I've got a second date.'

19

Inside

Days in Withered Hill: 3

Sophie's second attempt to escape from Withered Hill takes place just two days after her failed bid to navigate the woods on her first night in the village. Catherine finds her not long after dawn, lying in the dew-damp grass in the ragged remnants of her clothes, skin torn by brambles and hair knotted with burrs.

'Silly cow,' mutters Catherine, putting her hands under Sophie's armpits and unceremoniously hauling her to her feet. 'Still, I suppose you have to learn somehow.'

Catherine puts her in the bath and sponges her down, Sophie wincing as the soap gets in the scratches that criss-cross her flesh. Then Catherine patiently untangles the seeds and twigs from her hair and shampoos her. Sophie sits in the water numbly, calmed by Catherine's ministrations but not eased by them.

'Don't get used to this. I'm not a servant.'

Sophie glowers at Catherine, who is kneeling by the bath, her sleeves rolled up, her white shirt wet and clinging to her chest. 'What are you, then?'

'A teacher,' says Catherine, looking at her wristwatch. 'And I'm going to be late, especially as I have to get changed now.'

Catherine stands up and strips down to her white bra. She catches Sophie looking at her and meets her stare. Sophie looks away and Catherine stuffs her shirt into the laundry basket.

'I suppose you're going to try to leave again while I'm out. I wouldn't try the woods again. If you decide to stay, there's food in the fridge. Help yourself to anything.'

'Why am I a prisoner here?' says Sophie sullenly.

'You're not a prisoner. You just can't leave.'

'Is there a difference?'

Catherine looks down at her black trousers. 'Bloody fuck, they're soaking as well.' She unzips them and steps out of them, kicking them towards the laundry basket. She looks at Sophie. 'Yes, there is a difference. You just have to work out what it is.'

Sophie watches her stuff the trousers into the basket. She says, 'What were those things in the woods last night?'

Catherine doesn't look at her. 'Things?'

'People, then. Except not. I don't know. That's why I'm asking.'

Catherine is silent for a long moment, then says, 'Withered Hill isn't like other places. It's special. They… they're what make it special.' She looks at her watch. 'I really have to go.'

'What am I supposed to do while you're teaching?'

Catherine shrugs. 'If you're not going to try to leave, you could work out how to use the washing machine. It's not rocket science. You could even prepare us something for tea.'

Sophie stands up in the bath, the soapy water dripping off her, and takes the towel offered by Catherine, who appraises her mildly.

'Don't worry, I'm not going to pounce on you. Despite what I might or might not have said last night when I was pissed. I might be horny when I've had a drink but I'm not stupid. Get dried off and dressed; I'm going to work.'

Sophie wraps herself in fluffy white towels and goes into the bedroom as she hears the door slam. She stands at the window and watches Catherine walking over the meadow and towards the village. When Catherine has gone, Sophie looks towards the woods, where she feels eyes upon her, hidden in the dark depths, observing her with as much curiosity as she has with Catherine.

There are clothes left for Sophie. They are not new, but they fit well enough. She wonders who has worn them before. Nothing fancy; just jeans and plain T-shirts. She is starting to get a sense of herself, though where that has come from she doesn't know. It is as though memories are crowding at the boundaries of her mind. But memories of things she has no memory of. She shakes her head. She doesn't understand what she means. How can a memory not be a memory?

She wanders around the house, picking things up and staring at them. A candle. A teapot. A book. A shoe. A sharp-bladed kitchen knife. She knows what these things are, of course. Is that memory? Or knowledge? Are the two the same, or different? On a small table in the living room is a phone, connected to the wall. Beside it is a mobile phone. Catherine must have forgotten it. She picks it up and the screen lights up, displaying the time and date, overlaid on a picture of Catherine wearing a bikini on a beach, against a blue sky. So not everyone is a prisoner here in Withered Hill. Catherine at least can leave, or has done.

The phone is not locked, and Sophie swipes up to the keypad. She could call someone. She doesn't know who to call. She searches around for a number on something, eventually picking up the paperback book she touched earlier. A novel, a thriller, about a woman whose life may or may not be what she thinks it is. On the flyleaf are the details of the publisher, including a telephone number. Sophie punches it into the phone and presses the green button. She puts it to her ear, but there is just static, dead noise.

Sophie replaces the phone and picks up the receiver from the landline. She puts it to her ear and is about to dial the publisher's number again when a voice sounds. 'Hello? Withered Hill exchange.'

'I need to leave here,' says Sophie in an urgent whisper, though there is no one else in the house to hear her. 'Can I order... a taxi? Or something?'

The woman at the other end of the line laughs. 'Sophie Wickham! It's Carol. From the Post Office. Don't be a silly ninny.' Then the call goes dead.

–

When Catherine returns to the house at four o'clock, she pauses in the doorway to the kitchen where Sophie is washing up, sniffing the still air agreeably. 'That smells good.'

'Just a stew,' says Sophie. 'I found some things in the fridge and some vegetables and just threw it all in.'

Catherine looks impressed. 'If it tastes as good as it smells, you're a keeper.'

'I worked out how to do the washing machine,' says Sophie. She simultaneously wants to impress Catherine, and hates herself for doing so. She points through the kitchen window to the garden. 'It's all on the line.' Sophie wipes her hands on the tea towel and turns to Catherine. 'If I can't leave here, what am I supposed to do?'

'Work out how you *can* leave,' says Catherine, taking four bottles of wine from a tote bag and loading them into the fridge.

Sophie gives a frustrated sigh. 'Is everything a puzzle here?'

'See my car out front?'

Sophie nods. 'The yellow Mini.'

Catherine smiles tightly. 'Very good. Do you know how it works? How the engine makes it go?'

Sophie shakes her head.

'Me neither. It's a puzzle. But I could get someone in the village to show me. Terry, in the garage on the south road. I could learn what each part does, how it interacts with the others, how it burns fuel to make the car move. Then it wouldn't be a puzzle, would it? It would be knowledge.'

'So...' says Sophie carefully. 'I just have to find out how Withered Hill works, then I can leave?'

Catherine shrugs, and takes the last bottle out of the fridge that she'd put in. 'Maybe. That's what you'll have to work out. It's still lovely outside. Join me for a glass of wine in the garden?'

'You need some clothes of your own,' says Catherine, her words slurring slightly. They have had a bottle of wine in the garden, then moved to the kitchen to eat Sophie's stew, which Catherine enthused over, and are now sitting in the living room as the summer night has fallen, drinking.

'Who did these belong to?' asks Sophie, looking down at her jeans.

Catherine shrugs. 'What do you like?'

'I don't know. What do you think I should wear?'

Catherine, sitting on the sofa while Sophie is curled up on the chair, puts her head on one side and regards her critically for a moment. 'I'm not sure. Jeans suit you. Maybe some shirts. Vest tops. Shorts. It's going to be a warm summer.'

'How do you know?'

'It's always a warm summer in Withered Hill.'

Sophie takes a drink. This time, she is not faking it with water. She is allowing the wine to pleasantly fuzz her head. 'Where will you get them? I've not seen any shops in Withered Hill selling things like that.'

Catherine taps the side of her nose with one finger. 'Tomorrow's Saturday. I'm going outside.'

Sophie blinks. 'Outside Withered Hill?'

'Outside Withered Hill. I'll be setting off early.' She squints at her wristwatch. 'We should go to bed. I'll try not to wake you in the morning.'

Days in Withered Hill: 4

Catherine leaves the house while the sun is low, not bothering to check in on Sophie's closed bedroom door. The Mini is old and she has to pull out the choke to get the engine going, then bounces along the rough track to the main street, heading south. She beeps her horn, a thin, weedy sound, and there is a reply from a throatier vehicle.

Sophie waits a few minutes before lifting her head from beneath the two coats that were on the back seat when she sneaked in while it was still dark. She has folded herself into the tiny space in the footwells of the rear seats, and her legs are cramped from the tight, uncomfortable position she has maintained for forty minutes. As she straightens up, Catherine looks at her in the rear-view mirror and sighs.

They are approaching an arch of trees that overhangs the road, creating a green, leafy tunnel that connects Withered Hill to the outside world. Sophie reaches around the driver's seat and places the blade of the kitchen knife lightly against Catherine's throat.

'Keep driving.'

'You think this is going to work?'

Sophie nods at the road ahead. 'This isn't the woods. We can't get confused or turned around. There's a road. Keep driving along it.'

'You're not ready to leave,' says Catherine.

'I'll be the judge of that.'

Catherine shrugs and looks back at the road. The tree arch is a hundred yards away. Fifty. Twenty. Then the engine cuts out with a whine.

Sophie presses the edge of the knife against Catherine's neck. 'Stop that.'

'I didn't do anything.'

The Mini coasts to a halt, ten yards away from the arch. Beyond it, Sophie can see the road tantalisingly winding through the moors and fells. To the outside. She says, 'Start the engine.'

Catherine turns the key in the ignition. The car splutters and complains but the engine doesn't catch.

'Stop messing about. I'll do it, you know. I'll cut your throat.'

'You try, if you don't believe me. Can you drive?' Catherine sounds impatient.

Sophie hesitates. Yes, she thinks she can drive.

Catherine says. 'I'll get out. You get in the driver's seat. Look, take the car. Leave me here. Drive out of Withered Hill. Go.'

Sophie thinks, then nods. Catherine climbs out of the car and stands back, by the treeline, to show she isn't going to try anything. Sophie pushes the seat forward, gets out, then sits back down in the driver's seat. She points the knife threateningly at Catherine, who shrugs. Sophie turns the key in the ignition; the engine coughs, then dies. She repeats the operation three times, then gives up and gets out of the car. 'It's broken.'

'It's not,' says Catherine gently. 'It's you. You can't leave yet, Sophie. They won't let you.'

'When, then?' says Sophie fiercely. 'What do I have to do? Why am I here?'

Catherine gently takes the knife from her and throws it in the back seat of the car. She takes Sophie's hand and leads her to the side of the road. 'I know this is difficult, and confusing, but the sooner you accept it, the quicker you'll get to leave. It's just the way it is.'

'What are you going to do about the car?'

Catherine sits in the driver's seat, turns the key, and the engine growls into life. She shuts the door and winds down the window. 'I'll bring you something nice back from the outside.' She smiles.

Sophie stands by the side of the road and watches Catherine drive under and through the green arch, and onto the road beyond. She moves to the middle of the road and watches the yellow Mini receding until it is lost from sight behind a hill. On its way to... where? To outside. To everywhere.

Sophie turns around and looks up the hill towards the village. Towards home.

20

Inside

Days in Withered Hill: 360

Sophie's cottage, which she was led to a week after staying with Catherine when she first came to Withered Hill, is on the west side of the village, in the knot of dwellings behind the town hall and square. It is small, just one bedroom, but cosy, with a wood burner in the hearth and a tiny kitchen with a range that radiates intense heat in the winter, which is where she is standing, sipping coffee and staring out of the window at the small yard and the wildflower meadow beyond, when there is a smart rapping at her red-painted front door.

At first, she thinks it is Catherine, back from the outside, bringing whatever spoils she has triumphantly returned with this Saturday afternoon, but when she opens the door, it is Thaddeus Obermann standing there, dressed in a tweed jacket and plus fours, his feet shod in walking books and holding a stick with the carved head of a goat for a handle. He is leafing through a small, hardback book of some age and looks up at Sophie, blinking, as though he started reading it in his library and is not quite sure how he arrived at her door.

'Thaddeus,' says Sophie pleasantly. 'To what do I owe the pleasure? Do you want to come in? I've a pot of coffee on.'

He looks down critically at her shorts and then up to her vest. 'Oh, this won't do at all. Don't you have any better clothes?' He raises one eyebrow. 'Not that they aren't diverting garments in their own way.'

'Better clothes for what?' says Sophie. The sky is blue and cloudless and the sun is casting sharp, early-afternoon shadows. Her outfit seems eminently suitable for the June weather.

'We're going walking, of course. Through the woods.' He frowns, then smiles. 'It's time I showed you something.'

—

Sophie has assembled a wardrobe of clothing both new and inherited over the course of the last year, so she changes into a pair of jeans, throws a shirt over her vest, and puts on a pair of walking boots. Some of the clothes were hanging in the wardrobe of the cottage when she took possession of it, when it had begun to sink in that she was not leaving Withered Hill any time soon, and even if she could, she didn't know or remember enough about the world outside to know what to do or where to go.

'I'm not the first one to come here, am I?' she said as Noah, Carol and Catherine showed her around the cottage. 'There have been others. What happened to them?'

'They left,' replied Carol with an encouraging smile. 'When it was time. When they were ready. They all left.'

'That's more like it,' says Thaddeus when Sophie descends the narrow staircase into the living room in her new outfit.

The old librarian leads Sophie out of her cottage and across the street, cutting between two small houses and following a track that leads them to the edge of the thick woods. Sophie looks into the shadowed recesses where the track disappears. 'Are we going to meet... them?'

'Oh, good heavens, no,' says Thaddeus. He pauses. 'I mean, it's entirely possible. But I shouldn't expect so.'

'Will we be going to the edge?' asks Sophie. She wonders if this is some elaborate pantomime, and that Thaddeus is actually leading her out of the woods on the other side, out of Withered Hill. Has she passed the test without knowing it? Something flutters in her chest and simultaneously feels leaden in her stomach. Surely not. Surely when it happens, when she leaves, it will be with more ceremony than this.

Thaddeus lets out a rumbling laugh and taps his stick on the trunk of a venerable oak as they walk into the shadowed depths. On it is carved a heart, with two sets of initials inside. TO and CM. 'Carol Mountjoy and I,' he explains unnecessarily. 'We tumbled here on occasion as callow youths, up against this very tree.'

Sophie smiles to think of Thaddeus and Carol as young people. 'Did you never think to marry?' She remembers the term most people in Withered Hill use. 'To handfast?'

Thaddeus leans on the tree, tracing the outlines of the initials with fingers as gnarled as the bark. 'Ah, perhaps we talked about it, lying here in the midsummer night. But I went away to university. Manchester College, Oxford, to read Classics.'

Sophie's eyebrows rise. 'You left Withered Hill?'

'For quite a while. Fifteen years. We are not prisoners here, Sophie Wickham. Not even the likes of you. Surely you understand that, by now? We in Withered Hill are not your gaolers. We are your teachers, your guides. We are preparing you for the outside.'

'How many more like me have there been?'

Thaddeus starts to walk on, the path narrowing between the thick trees. 'A number.'

'And it takes them all a year? Before they are told they can leave?'

She sees his shoulders rise in a shrug as he leads the way. 'Some are here for a shorter time. Others for longer. It isn't an exact science.'

The way is becoming harder going, and Sophie is glad of her boots and jeans as Thaddeus brushes away with his stick the tangles of brambles on the path.

'And has anyone tried to escape as often as me?'

–

The next time Sophie tries to flee Withered Hill happens four days after she held a knife to Catherine's throat in her failed attempts to simply drive out of the village, which makes three times in her first week there.

She wakes at dawn every morning and stands in the window of her bedroom in Catherine's cottage. Sometimes she sees the feathering contrail of an aeroplane high in the sky, and thinks that might be the only way she can leave Withered Hill, to fly out of there. But that is impossible. So instead, Sophie opts for the improbable.

Although the village is surrounded by the impenetrable ring of trees, there is farmland belonging to Withered

Hill in the territory immediately outside. From Catherine's cottage, she sees, every morning, Peter O'Keeffe, the burly tattooed farmer she saw on her first day, and who sits on the parish council, driving his sheep along the south road and under the trees. It's a large flock, a mass of cloudy white that huddles together and skips along the road, marshalled by Peter and his dog. Tightly packed, thinks Sophie as she watches them. A sheep in the middle of that, it couldn't turn around even if it wanted to. Even if it was compelled to.

The next morning, Sophie awakes before dawn and in the darkness dresses in a white shirt and a pair of grey trousers, and wraps a white scarf around her dark hair. She steals out of the house as silently as possible and makes her way towards Nut Nan Farm.

The sheep are already gathered in the yard, at the gate, making their insistent baa noises, conditioned for their morning walk to the grazing fields outside the village. There are lights on in the farmhouse, and the sky is paling in the east. Sophie climbs over the gate and crouches low, pushing in between the jostling sheep, making her way to the centre of the flock. She isn't prepared for how bad they smell, nor how they gaze at her with a combination of vague curiosity and slight annoyance in their yellow eyes. Breathing through her mouth, Sophie gets down on all fours, in the muddy, shitty ground churned up by the sheep's cloven hooves, and puts her head down in between the two animals either side of her.

It seems like an age until she hears Peter O'Keeffe whistling and calling his dog, and then the gate is opened and the flock moves forward, Sophie crawling with them, pushing herself as low as she can.

The flock moves faster than it seems to do when she watches from the window, and she struggles to keep pace. Her hands are pricked and cut by the stones on the road, the knees of her jeans are wearing through and she feels the skin beneath them stinging. She is falling further behind in the flock, as well, and tries to redouble her pace, hoping the relentless voices of the sheep are drowning out her heavy breathing and occasional stifled cry as her flesh opens on another stone.

She daren't look up, but she senses they are approaching the trees. The air gets colder, thicker, somehow; she does not feel the morning sun on her back. There is something else, as well. A sense of more presences around her than just the sheep, Peter and his dog.

Peter calls, 'Come by!' and whistles a complicated volley at his dog. The sheep slow and seem to turn to the left, and Sophie turns with them. It feels dark; they are well in the trees now. Surely the road must lead out at any moment. Why have they stopped?

Then the sheep seem to quiet, and sink lower. They are sitting? But no, not sitting, at least not like she has seen sheep do before. They are lowering their front ends, laying their legs flat, bowing their heads, as if in silent supplication. Sophie pushes herself flat on her belly, feeling the earth beneath her at the side of the road.

'I would thee grant me a boon!' calls out Peter in his thick, strong voice. Who is he talking to? 'I would have safe passage to the grazing lands and the privilege of thy watchfulness on my flock, so they grow round and fat and bring us good fortune in Withered Hill.'

Sophie has her cheek against the dirt, trying not to breath, listening for some kind of response. There is none – at least, none she can hear.

Then Peter says, 'Of course, thy will have an offering to show my gratitude.'

She senses movement at the fringes of the flock behind her and holds her breath, closing her eyes tight. Then she is grabbed by the neck and gasps as she is hauled upwards to her knees. It is Peter, of course, and he is holding a long-bladed knife, more a machete, dull and pitted. He puts the blade against her neck and she looks around and up at him with fearful eyes. He grins savagely at her and turns to the thick, densely packed trees.

'This 'un? No, I thought not.' Peter lets her go and she tumbles forward, and he instead grabs the sheep next to her, pulling it up by the scruff of its neck until it lays against him. 'This one more to your tastes, aye?'

Then he draws the knife across its neck, splashing Sophie with the blood that gouts from its arteries. Wiping the blade on his trousers, he sheathes it at his belt, hefts up the still-twitching animal, and strides through the flock to the edge of the wood, tossing it into the undergrowth. The animal lies there, its yellow eyes meeting Sophie's, until the light in them slowly fades.

'I thank thee,' calls Peter into the dark woods, then turns to Sophie. 'And you should get home, Sophie Wickham, and get cleaned up.' He looks at her for a long moment. 'You're a pretty thing. Could put that energy to a lot better use than crawling around with sheep.'

Sophie gets to her feet and watches Peter drive his flock on through the trees, the outside tantalisingly close. Then she turns around for the third time that week and walks back to Withered Hill.

–

155

'Oh, there have been a few who have tried to escape many, many times,' says Thaddeus, striding on into the woods. The path has become lost now, and Sophie is trusting that he knows where he is going. 'But I imagine you're up there.'

'Thaddeus, where are we going?' asks Sophie as he swipes at a mass of brambles like an old-time jungle explorer.

'Here,' says Thaddeus, leading her into a sudden clearing illuminated by shafts of sunlight. It's not a natural glade, though; it is formed from carnage. The trees are broken, some of them dead, and the ground seems lower, like a trough of some kind, leading out from where they stand to the edge of the woods, so temptingly close. The outside is obscured by smaller, thinner trees, while the broken mighty oaks are coated with moss. Whatever happened here happened a long time ago.

'It was on this very day eighty years ago,' says Thaddeus in sombre tones. 'June the fifteenth, 1944. To the west of here, the United States Army Air Forces were allowed to operate from a British airbase to strike against the enemy forces in Europe. On the return journey from such a mission, a B-24 Liberator bomber crashed, right on the outskirts of Withered Hill. Its fuselage skidded many, many hundreds of yards, coming to rest here, right in our woods.' Thaddeus puts his hand on one of the splintered old oaks. 'Fire and cold iron. It was devastation for those who lived in the woods.' She knows what he means. The things... people she has seen. The dark, chittering, chattering things, that may or may not be human. He turns to look at Sophie. 'And that was the one and only day someone like you left us, before their time.'

21

Interlude

June 1944

It is a foggy day, Margaret notes with dissatisfaction. Quite the disappointment for flaming June, especially after the lovely weather they have had all week. The mist obscures the fells and curls around the woods that surround Withered Hill, feathery tendrils probing the outer ring of trees as though testing defences, but not venturing deeply inside. As a result, there's blue sky overhead, but the sun is still yet to rise above the fog. What is the point of a flawless sky without sunshine?

Margaret turns from the doorstep of her cottage and goes to the little pantry to decide what to have for luncheon. Rationing is not a great problem in Withered Hill, at least not for fruit, vegetables and meat. Well, most fruit. Elizabeth Mountjoy who runs the postal office has been craving bananas for most of the year, and cannot wait for this damnable war to be over for that very reason alone. It has been two weeks since what they are calling D–Day, and the Allied forces are pushing deeper and deeper into France. Perhaps, as everyone says, it might well all be done by Christmas.

Margaret turns on the wireless to listen to the news programme. They have started calling the new type of bomb that killed eight people in London two days ago the 'doodlebug', which Margaret thinks is an overly comical name for something so deadly. However, according to the newsreader, an RAF Mosquito has shot down one of the V-2 doodlebugs over the Channel, which is cause for great celebration.

There is a sudden banging and clattering at her door, and like a miniature whirlwind, Gladys powers into the cottage, laden with paper bags. She has been on one of her frequent visits outside, and she always procures something nice for Margaret.

'What's this boring thing doing on?' declares Gladys, glaring at the wireless. Margaret notes she is wearing new nylons that she wasn't wearing when she left Withered Hill yesterday. No doubt there will be risqué stories about American GIs over gin and lemon in The Farmer and Devil later.

'I am merely educating myself about the outside world,' protests Margaret. 'For when I leave Withered Hill and return to it.'

'Pshaw!' says Gladys, dumping her bags on the sofa. 'You're nowhere near ready to leave yet. Besides, you don't want to go out there. Not until the war's over and rationing has stopped. You wouldn't believe what I had to do to get half of this loot.'

I probably would, thinks Margaret, as Gladys bends over the wireless and turns the dial until she finds music.

'Oh, I adore this!' says Gladys. '"I'll Walk Alone". Dinah Shore.'

She starts to half-sing, half-hum the tune, and straightens up, taking Margaret's hands and pirouetting her around the small cottage living room. Margaret can't help but laugh.

'You are a silly sausage,' she says, breaking away. 'I was just about to make a pot of tea, if you're interested.'

'I think we should go and get a little libation outside the pub,' says Gladys. 'Show off your new frock.'

She delves into one of the bags and pulls out a navy-blue tea-dress with white polka dots.

'Oh, it's gorgeous!' says Margaret, holding it against her.

'And nylons,' adds Gladys, waving them at her. 'Don't ask how I got them. Or do, but after a wee drinkie.' She also pulls out a package of brown paper tied with string. 'And that new book you asked for.'

Margaret tears off the paper. '*Fair Stood The Wind For France* by H. E. Bates! Thank you, darling!'

'I took a look at the flyleaf in the shop,' Gladys says with a sniff. 'Quite why you want to read a book about the bloody war is beyond me.' She takes the book from Margaret and tosses it on the sofa. 'Now. Get into this dress and let's go and show ourselves off on Main Street.'

'Yes, but who to?' says Margaret, heading up the narrow stairs.

Main Street is busy, but with the very old and the very young. Most of the men have gone off to fight; Withered Hill has never shirked its duties in time of war. As Margaret understands it, which is not a great deal, not yet, Withered Hill's menfolk are afforded some kind of… protection. In the Great War, they formed a pals regiment, and served at Ypres, and every single one of them returned. So it is expected to be when this horrid war is over.

The womenfolk are tending the farms and maintaining Withered Hill's enviable reputation for its output of greens and meat. Margaret, by dint of her unusual standing in Withered Hill, is not expected to work the farms, though, out of guilt, and sometimes boredom, she has lent a hand on occasion. Neither is Gladys expected to get her hands dirty. She has a threefold role in Withered Hill, just like her mother before her, and her grandmother, and goodness knows how many generations back. She teaches at the school, she travels to the outside, and she takes care of people like Margaret. Doubtless, Gladys will take a husband soon – she is twenty-three, after all – or at least have one of Withered Hill's strong young men father her a child. And she will have a daughter, as her line always does, and that daughter will do the same jobs, and so it will go on. Forever, expects Margaret.

It was March when Margaret stumbled into Withered Hill, naked and dirty and without memory or knowledge. The parade of animal-masked villagers watched her, took her in and gave her a home. Margaret does not really understand why she is here, but she knows there is a life waiting for her outside to take up the reins of once more, and she understands that she cannot do that until she is ready, or until others decide she is ready at any rate. That is fine. Everyone has their own duty to fulfil during these trying times, and this – as unorthodox and unusual as it might seem – is Margaret's. The time will come when she will be allowed to leave. She has been told that repeatedly.

However, as she walks along Main Street with Gladys, both of them in new frocks and nylons, their hair done up by Gladys's expert hand, Margaret is feeling a little more troubled than usual. The world outside seems so big and inviting and – war notwithstanding – *exciting*. She has only been in Withered Hill for three months yet it feels so small and constricting. Sometimes she hears the drone of the

American aeroplanes flying low overhead and she longs for them to take her with them, to fly her out to explore the world she is hungry to see again.

It's the again that troubles Margaret. Gladys brings her newspapers and magazines from the outside, the *Daily Sketch* and *Picture Post*. She devours them, every column inch of news and gossip, every monochrome picture, and dreams of them, until she is no longer sure what she remembers or what she has simply learned.

Gladys leads them to a wooden table outside The Farmer and Devil and says, 'I might go for a cider, actually, to wet my whistle. You?'

'I'll have the same,' says Margaret.

The mist has not cleared from the surrounding landscape, but the sun has risen above it and it is pleasantly warm in the beer garden of the pub. The cider is cold and sharp, the product of Withered Hill's orchards from last season. Margaret sighs happily. Yes, she is keen to see the outside world, but for now, this will do very well.

Mr Jones, the ARP warden, is pushing his bicycle up the street and leans it on the low hedge that surrounds the pub, mopping his brow with a handkerchief. The Germans have bombed Manchester and Liverpool, but there is little chance of them attacking Withered Hill, nestled in the Lancashire fells, hidden away from the world. Still, Mr Jones takes his job very seriously. He tugs at his collar and says, 'My, that looks like a good way to spend a Thursday.'

'Cheers, Mr Jones!' says Gladys cheekily, raising her glass. 'Join us for one if you like?'

He shakes his head. 'On duty, aren't I. You never know when—'

Mr Jones pauses and cocks his head on one side.

After a moment, Margaret can hear it too: the distant drone of an aircraft engine. None of them breathe for a moment, then Mr Jones smiles.

'B-24 Liberator, if my ears don't deceive me,' he says. 'Yankee boys heading back to Warton airbase after sticking it to Jerry, I shouldn't wonder.'

They all look up and around, but the mist that sits on the hills and fells is too thick to see the bomber. They can hear it, though, getting louder. And something else.

Margaret says, 'Should it be making that curious sound?'

Mr Jones frowns. 'It does sound a little... laboured, doesn't it? Wonder if they've taken a hit? Might explain why they're a little off course. That and this damned fog. Hope they get back to Warton in one—'

Suddenly, the engine drone cuts out completely. Margaret is just about to ask if the plane has flown beyond their hearing when there is an almighty, terrific crash from somewhere to the west, and the fog lights up a sickly yellow far beyond the treeline.

'She's down!' shouts Mr Jones, throwing himself onto his bicycle and beginning to pedal down the street, blowing urgent blasts on his little whistle.

'Should we do something?' asks Margaret.

Gladys shakes her head. 'It's a tragedy, but if it's come down outside Withered Hill, I don't really see that there's...' Her voice trails off as she looks from the elevated position of the pub towards the woods out west. 'Oh my God.'

Margaret follows her stare and sees, in the distance, the trees at the outer edge of the woods seemingly buckle and sway, and then there's an enormous explosion and a ball of black and yellow flame erupts into the sky.

Gladys seems almost frozen to the spot, staring at the growing column of black smoke. She is shaking her head, as though shocked into silence. 'No, no, no, no, no.'

'They'll sort it,' says Margaret confidently, though she has no real idea who she means.

Gladys shakes her head again. 'You don't understand… the trees… the woods. They are Withered Hill, and Withered Hill is them. If they die, we die.'

Margaret stares at her, at the fervour in her eyes. Then Gladys blinks and seems to snap back into herself.

'Come on!' shouts Gladys, and Margaret follows her, running down Main Street and hiving off behind the square, past Margaret's cottage and to the edge of the woods, where Mr Jones and a dozen of the villagers are already there. Thick black smoke is pouring into the sky from the outside edge of the woods.

Mr Jones looks at them. 'Water! Whatever you've got. Buckets, bathtubs, anything!' He looks back at the flames, and when he turns to them again, there are tears in his eyes. 'The woods. The woods.'

Margaret's cottage is closest, so they fill two buckets and run back to the growing crowd that is pushing into the trees, hitting at flames with their jackets, stamping on burning bushes. The water is taken from them, and Gladys makes to follow Mr Jones into the woods. She turns and says, 'Go back home. Get more water.'

Margaret watches Gladys disappear into the trees for a moment, biting her lip. Then she looks around at the chaos, and follows her.

There is a sound in the woods that Margaret hasn't heard before. Not animals. Something else. Something anguished and in pain, a noise of fury. And she isn't quite sure she hears it with her ears. It seems to echo around her head. And she feels that pain, too, deep inside her chest. She feels that anger, as though it is her anger, her pain. Or perhaps that she is of it.

She has lost Gladys's trail and picks her way through the woods. She has walked here before, of course, when she first arrived, and tried to leave. Then she found it impossible. Now, with the shrieking of the very woods rattling around her mind, she finds that the going is easier, that there are fewer, if any, obstacles to her progress.

Far off to her left, she sees flames, the very trees on fire. Behind her, she can hear shouts. No, not shouts. Singing...? She frowns and looks back towards the village. They are singing at a time like this?

The sound rises, as though buoyed on the hot air. A lament. A lament for the trees. There is something in the air around her, not heat, not flames, not the cracking of wood. Something else. Something Margaret can almost taste. Pain and death and terror. The trees are afraid.

Or, what lurks in the trees is afraid.

And there is the fuselage of the plane, shorn of its wings and tail section, blazing in the woods, where it has come to rest after crashing somewhere on the fells outside Withered Hill.

Outside Withered Hill. Margaret is almost at the edge of the woods. The trees are thin. She can see the fog-bound fells. She can almost touch them. Mesmerised, she holds out a hand, as though the outside is beckoning and she is answering its call.

Then comes another call. Gladys, shouting her name. Brambles and weeds tangle listlessly around her ankles, but they are weakened, and she kicks them away. Margaret doesn't respond to Gladys, doesn't even turn to look.

Instead, she puts down her head and runs, runs through the trees and out of Withered Hill.

22

Outside

Days to Withered Hill: 8

Sophie spent Friday evening taking photographs of the star carved into the floorboards from every angle, and close-ups of the symbols, and ran them all through Google image search. There were lots of matches, or near-matches, that took her to various occult websites, but nothing exact enough to shed more light on the star, other than it was likely used for protection. But protection against what? And was she the one being protected? Or were the symbols being used against her?

She'd decided not to contact the landlord about it, mainly because he'd no doubt make her pay for the damage, and she couldn't afford that at the moment. In the end, she just covered it up with the rug and poured herself a glass of wine. She didn't even believe in any of that stuff anyway. Did she? She sat on the sofa and scrolled through the images on her phone, pinching some of them bigger to inspect the symbols. One of her searches had taken her to YouTube and a video someone had uploaded of an old film called *The Devil Rides Out*. Sophie streamed it onto her TV and settled down with her wine.

It was pretty scary in parts, for an old movie. She sat up and paused it when the man who used to play Dracula and the guy who was in the old Prime Minister comedy show got into the engraved circle with another man and a woman, and lit big spooky candles all around it.

'Don't leave the circle!' shouted the Dracula man when a giant spider appeared ('Thanks for the trigger warning,' muttered Sophie, trying not to look at the close-up shots of its hairy legs and gaping mouth) and the little girl came into the room.

The pentagram on the floor in the movie was a lot neater than hers and had what looked like Latin engraved around the edges. Probably just some random stuff put together by the props department. Sophie rolled the rug back again and looked at the carving. She stood up and gingerly walked to the centre of it. Did she feel any different? Did she feel *protected*? She felt a sudden chill and rubbed the goosebumps on her bare arms, and laughed at herself. The sound was hollow and leaden in the still air of her flat.

She rotated slowly where she stood, looking from the centre of the pentagram out at her room. On the mantelpiece sat the corn dolly. Hadn't Colin said that was for protection, too? So whoever had done this and – because she couldn't find any record of actually buying it in her emails – sent the corn dolly too, were they actually trying to help her in some way? Protect her? But from what?

'It's nice to be asked first,' she said out loud. It didn't take away the fact that someone had been in her flat.

Sophie picked up her glass and went into the kitchen to open a fresh bottle. The sink was piled with dirty pots; she hadn't even bothered loading the dishwasher since she got in with Tom last night. She sighed and made a deal with herself: do the dishwasher and she could open another bottle, with it being Friday night. It was a hollow deal;

she'd open the bottle whether or not. But it made her feel like she was working for it, at least.

She was putting the cups and plates into the trays when her hand closed around her big carving knife, at the bottom of the sink. She hadn't even remembered using that. Frowning, she held it up to the light. The blade was covered with a fine, beige dust and the point was clogged with what looked like... pulp of some kind. She peered closer. No, fibres. Wood fibres.

Exactly the same wood as the boards in the living room.

—

Colin, it was fair to say, was surprised to hear from Sophie at almost midnight on Friday. She had all her colleagues' numbers and addresses in a little booklet she'd been given on her first day, including people at other offices who she was unlikely to ever meet, and which she never thought she'd ever need to use. Sophie messaged Colin from the sofa, the knife on the coffee table.

> Hey sorry its late its Sophie from work can I send you some pics?

Immediately, she realised how that sounded and the colour drained from her face. The three wavy dots indicated Colin was already replying, so she hurriedly typed again.

> Not that sort lol pics of something a bit weird I've found in my flat

> Hello, Sophie. What a nice surprise to hear from you. Of course you can, if I can be of any help I'd be delighted to take a look.

She smiled at his formality. And use of punctuation; full stops at the end of sentences and everything. She attached five pics to the message: one big one of the star and some close-ups of the symbols.

> Can you take a look at these? Let me know if they mean anything x

She grimaced at the unconscious kiss at the end of her message. She was already regretting contacting Colin. Too impulsive by half, after a drink.

> Of course, Sophie. I'll do some research straight away. Have you just found this under a carpet or something like that?

> Yeah, I took the carpet up cos I fancied bare floorboards and there it was, prob been there years. K, won't keep you, its late, thx again x

Aaargh, there she was again with the kiss. She could see him replying, but she flicked off the app and tossed her phone onto the couch. It beeped to indicate Colin's message, but she ignored it. She emptied the last of the second bottle into her glass and hit play on her phone to finish off *The Devil Rides Out*. But she couldn't concentrate. The knife sat on the table in front of her. The knife that had been used to carve the pentagram in her floor. The knife that meant it had been done in the past twenty-four hours or so. Which meant one of two things. Either she'd done it herself, while pissed, and couldn't remember, which even for Sophie was very unlikely, or it had been done by the only other person who had been in her flat. Tom Gisburn.

–

Sophie opened one groggy eye, sunlight flooding into her bedroom. The buzzer to her flat was ringing insistently. Sophie groaned and slid out of bed, looking down at herself. When had she put the black silk nighty on? She didn't remember doing that before bed.

She dragged her dressing gown on and shuffled to the intercom, yelling, 'All right! All right! I'm coming!'

It was probably a delivery. She'd no doubt ordered more clothes she couldn't afford last night.

She hit the intercom button and said, 'Yeah?'

'It's me. Colin. I've got something you should see.'

Sophie sighed and buzzed him up. She caught sight of herself in the mirror above the fireplace; her hair was all mussed and her face puffy. She pulled the dressing gown tighter around her as Colin rapped sharply on her door.

'You should have called first,' she said as she let him in. He was wearing knee-length shorts and a striped polo shirt tucked into them. Long white sports socks and cheap trainers. 'Have you found something out about the pentagram? What time is it anyway?'

'Sorry, I was just so excited to see you,' he said, brushing his fringe out of his eyes. And then she saw that, yes, he *was* excited to see her. She gaped at his shorts, then up at his face. 'Colin? I mean, what the fuck?'

Then he was on her, kissing her neck, turning her and pushing her back against the closed door, pressing his body against hers. She balled her fists to hit his chest, but he was stronger than he looked, pushing her hands back against the door and crossing her wrists, holding them in one hand while he fumbled at her dressing-gown belt. And suddenly she wasn't fighting him any more, she was dragging at the buttons of his shorts and—

'Oh my fucking God,' said Sophie, sitting bolt upright in bed. It was still dark, only the faintest glimmer of dawn glowing through the open curtains. She felt sick. In fact, she was going to *be* sick. Her head pounding, she swung her legs onto the carpet and just made it to the bathroom before she let loose a stream of hot, acidic puke into the toilet bowl.

She sat there for a long time, head against the cool rim of the toilet. Was she running a temperature? God, she didn't want to get ill. Not with her date with Tom on Saturday night. She eventually dragged herself to her feet and staggered to the kitchen, where she downed two pint glasses of water.

In the living room, the rug was still rolled back, displaying the carving. Fucking nightmares. She shouldn't have left that thing exposed, shouldn't have watched that stupid horror film, certainly shouldn't have messaged Colin.

It was only three fifteen. She filled her glass with more water and found a couple of ibuprofen, then got back into bed. Thank God it was Saturday. She scrolled through her social media for a bit, then put her phone face down on the bedside table and closed her eyes again.

She dreamed of Colin again, instantly. It was almost a direct continuation of the first dream, and even in her stupor, she found herself thinking, *I didn't know that could happen.* Now they were in bed, Colin saying over and over again, 'It's for your protection, it's for your protection, it's for your protection.'

'Fuck.' Sophie sat with her head in her hands in bed, then reached for her phone. She'd only been asleep for ten minutes. 'Fuck, fuck, fuck!' she said loudly.

Why had she messaged Colin? It was like she'd given him a hotline into her head. She didn't want to ponder too deeply what kind of fucked-up mess her subconscious mind was to dream about them doing *that.*

Sophie lay back on the pillow, afraid to close her eyes. She felt sick to her stomach again. To stave off sleep, she tried to psychoanalyse herself. Was she feeling angry at Jamie, because she was sure he could be the only person who had carved that pentagram? And she was... what? Projecting that anger and fear onto Colin, who – it had to be said – had only ever shown her kindness? And was this the only way she knew how to be nice to someone? To sleep with them? Was it some weird, screwed-up revenge she was taking on Jamie in her sleep?

. Sophie rolled onto her side, looked at her phone again and sighed. She couldn't stay awake all night. But she didn't dare go to sleep again. 'Don't leave the circle!' the Dracula guy shouted in her head as her eyes drooped. 'It's for your protection!' said Colin, his face contorted, looking down on her.

She jumped, wide awake again. 'Fuck this,' she said, getting out of bed and gathering up her duvet and pillow.

Then she went into the living room, wrapped the duvet around her like a sleeping bag, and lay down in the centre of the pentagram, where she had six hours of uninterrupted, dreamless sleep.

23

Outside

Days to Withered Hill: 7

Sophie felt weirdly cleansed from her sleep in the centre of the pentagram, save for a crick in her neck where she'd slid off the pillow. It motivated her to do some wider housekeeping, washing her sheets and cleaning the bathroom, polishing the surfaces and hoovering the rugs. The postman came with a stack of bills and late-payment letters, which she shoved into her knicker drawer, unopened, and there was a delivery of a new dress, a short black number, with some new underwear as well. The dress she'd bought for her first date with Tom was balled up in the laundry basket; there was a stain on it, and a rip under the arm. Well, she couldn't wear that again for him anyway. She pushed it into the bin in the kitchen, then tied the bag and hauled it downstairs to the wheelie bins outside the front door.

As she turned from the bins, she caught a flash of something across the road, suddenly obscured by a passing bus. She didn't know what; a person? She didn't even know why they had caught her eye; some familiarity, perhaps? But when the bus had gone, there was no one she recognised, just an old lady being pulled along by three small dogs on leads.

Back in the flat, Sophie regarded the carved pentagram critically. In the warm June sunshine flooding in through the freshly cleaned windows, it all seemed a little... silly. She inspected the knife again. It could be wood fibres on the end. It could equally be... She shrugged. Bits of an avocado stone. Anything.

And something else had started to form in her head. An idea she didn't want to entertain.

What if it *was* Tom?

But surely she would have heard him carving shapes and stars into her floor. She'd not had anyone in the flat for... well, for weeks. Months. Apart from him. She knew, deep down, that she did not *want* it to be Tom who had done this. She did not believe it was Tom. She didn't know why it was there, or how, but it wasn't him. And tonight she was going to prove that to herself.

Her phone buzzed, and she jumped. It was a call; her gran. Sophie felt a pang of guilt; she hadn't called her for ages.

'Gran!'

'Sophie? Is that you?' said her gran cautiously. Sophie didn't know who else she thought might be picking up the call.

'I'm sorry I haven't called. I've just been so busy at work. I got a promotion. Well, a full-time contract. And they've been working me really hard.'

'They're building houses on the field,' said her gran. 'Affordable housing. Whatever that means. Everything's affordable if you can afford it, isn't it?'

Sophie frowned. 'OK. That's a good thing, isn't it? People need places to live. And nobody young can get on the property ladder in the village.'

'That's the field where we used to hold the fetes. Before your time. Before your mum's time. Oh, yes. That's why I'm ringing.'

'Because of the field?' said Sophie, a little lost. 'I mean, there are plenty of fields round there…'.

'Because of your mum. And your dad. You do know what day it is, don't you?'

Sophie opened her mouth to speak and then remembered what the date was.

–

'I still can't believe it,' said her father for what must have been the hundredth time.

'Brian,' said her mother, mildly but warningly from the passenger seat. 'This can wait until we get home. Concentrate on the road.'

Sophie stared out of the window from the back seat. Or tried to. It was dark and all she could see was her face reflected back at her, the short, messily cropped hair, the piercings in her nose and lip. She had panda eyes from where her mascara had run when her father had shouted at her relentlessly while he loaded her bags and boxes into the boot of the car outside the halls of residence. She should be going home for Christmas. Instead she was going home for good.

'Jean, can you imagine if this had gone any further? If the police had become involved? If it had gone to court?' Her father was seething, hunched over the steering wheel. He kept staring at her through the rear-view mirror, but Sophie refused to meet his furious gaze.

'It didn't,' said her mother calmly. 'It got dealt with well before it got to that stage.'

'If by "dealt with" you mean our daughter being thrown out of university, then fine, good, yes, I'm so glad it's bloody *dealt with*.'

'Brian. Calm down.'

'Don't tell me to calm down. Jesus bloody Christ, have they never heard of street lights around here?'

'We're not on a street, we're on a country lane,' muttered Sophie.

'Ah!' said her father triumphantly. 'It speaks! Well, while you're in a talkative mood, maybe you'd be so good as to start explaining yourself.'

Sophie tried to look past her ghostly reflection at the dark, treelined road whizzing past outside the car.

'You've ruined your life, you know,' said her father, glaring at her again in the mirror. 'Things like this, they're hard to keep secret. No one will trust you again.'

'Brian,' said her mother quietly. 'Please just concentrate on the road.'

'She'll never find a bloody husband,' her father remarked shrilly.

'What's that got to do with it?' shouted her mother back.

'Jesus Christ,' muttered Sophie. 'It's not the nineteenth century. You don't have to marry me off.'

'Nobody would bloody have you!' shouted her father. 'What I wouldn't give for someone to take you off my bloody hands, so I never had to see you again!'

The radio was playing Christmas songs relentlessly. They had never sounded so hollow, so false, so wrong. Her dad furiously fiddled with the dial.

'Brian!' said her mother sharply.

'Don't Brian me!'

'Brian!' her mother screamed. 'Stop!'

Sophie just had time to look ahead, where, in the middle of the road, strangely illuminated in the otherwise pitch-blackness, a figure was rising up, a man cloaked in black rags, like a crow, his face preternaturally pale, rising as though growing out of the tarmac, or like a balloon inflating, arms straightening as they sped closer and closer. Then the car flipped and everything turned upside down. Everyone was screaming and it all went black.

Sophie awoke briefly, every inch of her body shrieking in pain, her face in wet grass. She closed her eyes. When she opened them again, there was a riot of activity, two paramedics leaning over her, the treelined road painted by flashing blue lights. Just before they injected her with morphine and she lost all sense of anything once more, her eyes fell upon the blazing, upside-down wreck of her parents' car.

–

'I hadn't forgotten,' said Sophie down the phone to her gran, though it burned her cheeks with guilt to realise that she had.

'Will you go to their grave?' said Gran.

'Perhaps next weekend,' said Sophie. 'I scattered some of their ashes here, remember. I'll go and put some flowers down.'

'I could go with you if you come up here,' pressed Gran. 'We could visit Emily's grave as well. You never go to Emily's grave. She was your sister.'

'I'll call you in the week, Gran.' Sophie killed the call.

Sophie ran a bath and luxuriated in the bubbles for an hour, topping up the hot water when it cooled. She lay back and covered her face with a flannel. She did not want to think about her parents, or Emily, or that girl at university. So instead she elected to think about Tom Gisburn. He would be expecting to sleep with her again, now. After they had on the first date. She wanted to. She was hungry for it again. But she had to be sure of something first.

When her fingertips had shrivelled like prunes, she got out of the bath and dried herself off, then spent another hour getting ready. She studied herself critically in the full-length mirror in just her new black underwear, before stepping into the dress and zipping it up. She wished she'd bought new shoes as well. Sophie opened her wardrobe and regarded the piles of shoes there. Maybe three dozen pairs. Maybe more. Half of them black. She didn't need new shoes. She *wanted* new shoes, though.

Tom was picking her up at seven. At six thirty, she poured herself a large glass of wine and sat on the couch, waiting with butterflies in her stomach until, right on the dot of seven, her phone buzzed with a text from Tom saying *Outside now if you're ready.*

Quick change of plan, she texted back. *Can you come up for a minute?*

24

Inside

Days in Withered Hill: 238

In Withered Hill, they celebrated Faunus in February, on the thirteenth day.

'Strictly speaking, there were historically two festivals with that name, one on December the fifth,' Mr Obermann tells Sophie one cold, dark day in the library, as they huddle around the wood burner drinking hot chocolate laced with brandy while the snow piles up in the streets outside. 'But we adopted the February one. As you've already seen, we like to start Yule early. Two big celebrations in December...' Mr Obermann pats his gut, and laughs. 'I think it would probably kill us off.'

Mr Obermann hands Sophie the big book that sat on his lap and she looks at the pictures of the goat-faced man with shaggy hindquarters and cloven hooves, sometimes draped in foliage, other times naked and standing proud.

'In many ways, Faunalia is the most important of Withered Hill's festivals,' he says. Sophie hands him back the book. 'Are you sure you read this properly? It's quite important. The festival is in honour of Faunus, the Roman god who dwelt in the woods and was worshipped by farmers in return for bountiful harvests. But it is not just the fertility of our land and animals that we ask for at Faunalia. Do you know when the most popular month for birthdays is in Withered Hill?'

Sophie shakes her head.

'November. We are almost all Faunalia babies here. My own birthday is December the fifth, and I feel especially blessed. Conceived on one Faunalia, born on the other.' He winks at Sophie. 'But I'm not the old goat who gets it this Faunalia.'

'What do you mean?'

'You shall see, Sophie. You shall see.'

Days in Withered Hill: 241

Two days before Faunalia, Catherine lets herself into Sophie's cottage without knocking. She has been outside, and has come back laden with bags. 'Got you something to wear for the festival,' she says.

Sophie strips to her underwear in the living room and pulls on the brown woollen dress that fits her like a glove and comes just to her knees. She snuggles into the loose roll neck. 'So soft!' she says.

'Cashmere. Got it in the sales.' Catherine opens another bag and pulls out a pair of knee-length black boots with a small, square heel. 'Try these.'

Sophie puts them on and does a little twirl.

Catherine watches her with her head cocked to one side. 'Very nice.'

Sophie goes to make tea and calls from the kitchen, 'So what exactly happens at Faunalia?'

'There's a big fire, of course. There's always a big fire. There'll be a sacrifice. We take a goat down to the woods and slit its throat.'

'Ah, that's what Mr Obermann was on about, then,' says Sophie.

Catherine frowns. 'He told you what happens?'

'Not really. Just said something about him not being the old goat that gets it this year. Something like that, anyway. Is that it?'

'Oh, no. Then there's music and dancing, drinking and...'

Sophie carries the tray of tea into the living room. 'And?'

Catherine shrugs. 'Shagging. Lots and lots of shagging.'

'And is that mandatory?'

Catherine laughs. 'Sophie Wickham is no prude, so don't come to me with that. Not from what I've read on the socials, anyway.' She peers at her with narrow eyes. 'Are you blushing?'

Sophie throws a ball of tissue paper from inside one of the boots at her. 'No, I'm not. I just wondered.'

'It's not mandatory at all.' Catherine shrugs. 'It's just that everybody does. Mainly everybody. Most people round here, Faunalia was their first time. Ask some of the oldies, who can barely get out of their armchairs most of the year, and they'll always have a bit of lead in their pencils on February thirteenth.'

They sip their tea for a while and Catherine says, 'You haven't while you've been in Withered Hill, have you? Not as far as I know, anyway.'

Sophie shakes her head tightly. 'I thought it was best to concentrate on... you know, getting out of here. Rather than making connections.'

Catherine puts the back of her hand across her forehead in a fake swoon. 'I'm cut to the quick. You've put a knife in my heart. Not making connections?'

Sophie finds another ball of tissue and throws it at her. 'Not you. I meant… men.'

Catherine looks at her for a long time over the rim of her teacup. Eventually she says, 'Got your eye on anyone?'

'Peter O'Keeffe called round earlier,' says Sophie, by way of what she thinks is changing the conversation. 'To drop off some chicken and lamb for the freezer.'

Catherine raises one eyebrow but says nothing.

'I mean, he's married!' blusters Sophie. 'I couldn't! Not with a married man.'

'It's the tattoos, isn't it?' says Catherine, smiling.

'I just couldn't, it wouldn't be right.'

'There's no right and wrong on Faunalia.' Catherine puts her cup down on the coffee table. 'For one night only, all bets are off. Anything goes.' She stands up and gathers her bags. 'Thanks for the tea. Fancy a drink in the pub later? You need to get in training for the night after tomorrow.'

'Sure,' says Sophie with a smile, watching Catherine leave singing the song from that old musical, *Anything Goes*.

Days in Withered Hill: 243

It is a clear, cold night on the thirteenth, and after Catherine calls for Sophie, they walk to the town square, where the villagers are gathering. The adults, that is; anyone under sixteen has been locked away for the night, as is traditional on Faunalia, a cadre of the oldest of Withered Hill's residents who do not wish to participate fully in the festival travelling around the houses to visit and keep watch over them, and deliver them chewy cakes and hot chocolate. There are several braziers dotted around the square, and lines form behind them, the villagers each given a blazing, tar-capped torch.

Sophie is glad of the cashmere dress that keeps her warm on the freezing, luminous night, snow still clinging to the streets and frosting the trees. But the cold is tempered by a strange warmth emanating from the growing crowd. A hungry heat. Sophie catches glances cast in her direction, that linger too long, and in the dancing flames of the torches, everyone looks different than they do in daylight. There's a devilry about them, a musk that rises on the warm air of the massed torches, a sense of something impetuous and dangerous and barely contained that both frightens and thrills Sophie at the same time.

Noah Jones gives a speech about fertility and growth and the long traditions of Withered Hill, and then Peter O'Keeffe, still in shirtsleeves despite the cold, wrangles a goat on a short length of rope to the centre of the circle. He hands the rope to Zachary Winterbottom, who is wearing his best suit, shoes polished.

'To the woods!' calls Noah, and he marches off towards Main Street, Zachary falling in beside him with the goat, and everyone filing slowly up the cobbled road, their torches held aloft.

'It's going to be wild tonight, I can feel it,' whispers Catherine, her breath hot on Sophie's ear.

'Don't leave me,' blurts out Sophie, then feels stupid. She can't get in the way of Catherine's Faunalia; she can't expect to be babysat while it all goes on around her. She wonders if she can steal away at some point, just hide in her cottage until it's all over.

Catherine just squeezes her hand encouragingly, and they pick up the pace to catch up with the main procession now wending its way out of the village and towards the dark mass of trees huddled against the starry sky.

At the edge of the woods, Noah holds up his torch and the crowd grows silent. He calls out clearly, 'And now we give our gift. Step forward, Sophie Wickham.'

Sophie blanches and glances at Catherine, who gives her an encouraging smile. 'What's this?' she whispers fiercely. 'You could have told me.'

'It's tradition,' says Catherine.

'A *gift*? I'm the gift?'

Sophie has been in Withered Hill for just shy of eight months. Is this what she has been prepared for? Is this why she is here? Is this why she came to Withered Hill? She suddenly feels very afraid, a yawning chasm opening up in her knotted stomach. She remembers the drawings in Mr Obermann's book, illustrations of Faunus, his shaggy hindquarters, his huge, swollen genitals.

'Sophie Wickham!' calls Noah again, and she feels herself propelled forward by the crowd until she is pushed to the front where Noah and Zachary await. She looks around for Catherine, but she is lost in the sea of flame-painted faces.

Noah points at the dark woods ahead of them, and Zachary hands her the rope attached to the goat's neck. He also gives her a bone-handled knife with a serrated blade.

'You want me to go in there?' she says. Memories of her flight through the woods on her first night surface, the branches and thorns ripping at her clothes, the black shapes that chattered her name.

'You came to us from the woods, by Owd Hob's will,' intones Noah. 'On Faunalia, you must return.'

Sophie looks around, wonders if she can run. But where to? She has not tried to escape for a few weeks now, not since Yule Eve. She looks at the knife in her hand. There are too many of them, surrounding her. There is nowhere to go.

Except into the woods.

'You won't be alone,' says Zachary. 'I'll be with you.'

Sophie is pathetically grateful. She stands there, holding the goat, while the villagers crowd around Zachary, patting him on the shoulders and hugging him.

'Enough!' calls the old man. 'There's work to be done. Come along, Sophie Wickham.'

Swallowing drily, Sophie pulls on the rope and with one more glance back to try to find Catherine's face, and failing, she takes a deep breath and walks across the undisturbed snow towards the skeletal fingers of the trees reaching into the night sky, the goat trotting meekly beside her, Zachary on her other side.

When they get to the edge of the treeline, she looks back one more time. The villagers are touching their torches to a pyramid of wood in the clearing, and the bonfire is bursting into dancing flames. 'What now?' she says.

'We go in a little way,' explains Zachary. He cups his hands together and blows into them. 'Come on. We'll catch our deaths.'

Sophie follows him into the woods. Another look back and the trees seem to have closed around her, blocking any view of the villagers, save for a faint orange glow. She keeps walking through the virgin snow, stepping into Zachary's footprints, the goat sometimes stopping to nibble on a bit of ivy, other times pulling ahead of her.

A sudden breeze whispers through the dry branches.

It says her name.

Sophie Wickham, it says.

The gift, it says.

Sophie brandishes the knife out in front of her, slowly turning in a circle.

The gift, says the breeze.

'It's all right,' says Zachary softly.

The goat bleats. Sophie looks down at it. It fixes her with its yellow eyes, puts its horned head on one side. She remembers Mr Obermann asking her if she'd read the book properly, how important it was. She wracks her brains, trying to recall anything other than the grotesque drawings of Faunus. There was something about a goat, wasn't there? A sacrifice. She looks at the knife in her hand.

'I have to kill it?'

'Aye.'

'I'm sorry,' she says, as she slits the goat's throat.

The breeze intensifies, turns into a wind, whirls around her, whipping her hair, then it settles and fades. She looks down at the goat, its blood staining the pure snow. She has delivered the gift.

She looks at Zachary. 'Now what? Are we done?'

'Not quite.' He nods ahead, and for the first time, Sophie sees a dark shape just off the path. A rectangular hole, recently dug, the black earth piled to one side.

'Now I have to bury the bloody thing?' says Sophie.

'Not quite,' says Zachary again. He steps off the path and then, to Sophie's horror, sits down in the snow and begins to lower himself into the hole.

No, not hole.

Grave.

'What are you doing?' she whispers.

Zachary is smiling at her, his face bright in the moon-light. His eyes shining. 'I'm the gift, Sophie. Not that stupid animal. It's me.'

She looks at the bloodied blade in her hand. 'Then why…?'

'To bring them here, to let them know it's time.' Zachary suddenly disappears from sight.

Sophie walks over to the grave, looking down. Zachary is lying on the snow-flecked earth, smiling up at her.

'It's time for me to go back. Back to the earth. Back to the soil.'

Sophie flings the knife away from her. 'For God's sake, get up! Stop trying to scare me.'

Zachary doesn't move. 'I'm not. This is the way of things. Faunalia is about fecundity and birth and the planting of new life. But you can't have that without death. It's the cycle. It's the way of things.' Zachary continues to smile up at her. 'Fill me in.'

Sophie shakes her head, feels the panic rising within her. Is this some kind of test? To see what she'll do?

'Please,' says Zachary. 'It's what I want.'

'How can it be what you want?' says Sophie, her voice trembling.

'It's my time. I've lived my life, and lived it well. I'm ready to go, and be a gift to my kin, give them the harvest they deserve this year.'

'I won't do it.'

Zachary shrugs, then, with some effort, sits up. He reaches up to the edge of the grave and begins to pull the dark earth on top of him. Then he starts to sing.

'Oh Withered Hill, old Withered Hill, majestic and sublime, the praises we shall sing, they ring, until the end of time.'

'Please stop,' begs Sophie, her eyes prickling with tears.

'Thy beauty is eternal, thy banner flies unfurled,' sings Zachary in his croaking voice as he pulls more earth down onto his legs. 'Th'art the dearest and the grandest old hill in all the world.' Zachary looks at her. 'I can't do the next bit alone, Sophie. Please. It's the way—'

'The way of things,' says Sophie, her voice cracking.

'You'd shoot a horse when it was its time, or a dog,' says Zachary. 'This is no different.'

'It's what you want? Truly?'

Zachary nods. 'It's what I want.'

He lies back and starts to sing again.

'Withered Hill, old Withered Hill, thou standest quite alone. Twixt Burnley and Clitheroe, Whalley and Colne. Where Hodder and Ribble's fair waters meet, with hills and dales content at your feet.'

Sophie sinks to her knees beside the grave, tears freezing on her cold cheeks. Then, with a stifled sob, she puts her hands against the pile of earth, and pushes, gasping as she hears it land on Zachary in the hole.

'When witches fly out on a dark rainy night. We'll not tell a soul, and we'll bar the door tight,' Zachary sings.

Sophie scrunches her eyes tight, as though that will block out the awful sound, and pushes at the earth, hard and faster.

'We'll sit near to t' fire, and keep ourselves warm. Until once again we can walk on thy arm.'

She pushes and pushes, sobbing and sobbing. *It's what he wants, it's what he wants*. The words flit around her head like a murmuration of starlings.

Zachary continues to sing, and Sophie continues to push earth on him, until, mid-chorus, his voice is suddenly muffled, then silenced. Sophie pushes the earth until the hole is filled, then runs back to the village, heaving sobs ripping from her with every step.

Faunalia is in full heat when she gets there. There is music, and dancing, and by the light of the blazing bonfire, she sees people pairing off, heading into the edges of the forest. She hears the moans and cries of pleasure, sees the couplings against the trees, legs wrapped around waists, a thick, heady scent on the cold air.

Peter O'Keeffe is waiting for her by the bonfire. He takes the knife from her hand and wipes it on his jeans. 'You did it. Well done.'

'What did I do? What exactly did I just do? It was fucking barbaric.'

Peter shrugs, not taking his eyes off her. 'It's the way of things in Withered Hill.'

Sophie, exhausted, her hands filthy, looks around at the bodies painted by firelight and shadow. 'And this is the way of things, too?'

'Aye.' He moves closer to her.

Sophie looks around for his wife, Jan, and as though reading her mind, Peter says, 'She's off with Joe from Clough Hole Farm.' He smiles. 'Another Faunalia tradition.' Then he reaches for her hand and encloses it in hers, and draws her towards him. Sophie is mesmerised, like prey frozen in the stare of a snake. 'I want you,' he says.

'Do you want him, Sophie?'

She blinks, the spell suddenly broken, and turns around. Catherine is standing there, arms crossed, smiling.

'Well, do you?'

Sophie looks back at Peter. 'No. I don't. Not after what I've just been through.'

'That's all right,' says Peter good-naturedly.

'He'll not be short of a willing tumble tonight,' says Catherine.

Peter nods and walks away as Catherine joins Sophie at her side.

Sophie looks at her. 'And you? Have you had your Faunalia fun?'

'Not yet,' says Catherine. 'You sound angry.'

'I just buried Zachary Winterbottom alive,' hisses Sophie, almost unable to actually believe it herself.

'It's the—'

Sophie puts up her filthy hands. 'No. Stop. If one more person says *it's the way of things in Withered Hill* I'll… I'll…'

Catherine puts her hand against Sophie's and entwines her fingers around hers. 'What will you do?'

'I don't know,' says Sophie. Her breath catches in her throat as Catherine leans forward and gently kisses her neck. 'Is this allowed?'

'It's Faunalia,' says Catherine. 'Anything goes.'

Then she stands up on her tiptoes and kisses Sophie full and hard on the lips, then leads her by the hand back towards the dark, inviting woods.

25

Inside

Days in Withered Hill: 361

Sophie has been pondering the story Thaddeus Obermann told her, about Margaret, who was just like her, but
who fled Withered Hill before her time, in the confusion
of the plane crash. She imagines it like a fledgling falling
from the nest before it can fly, flapping around on the
ground, vulnerable to cats and rats. She wonders what
happened to Margaret. Whether she survived out there.
Sophie has begun to fear the outside almost as much as
she aches for it. She has a life waiting for her, but one
she can still only conjure in snatches and fragments, aided
by the social media posts and the documents that Catherine and others have presented to her. Birth certificates.
Newspaper clippings about the untimely death of a baby,
about a car crash, about the attempted suicide of a university student. And one that Sophie cherishes, protected
within the pages of a book. Only a small thing, a few
paragraphs, with a headline that says *Local girl graduates after
family tragedy*. There is a grainy photograph of a girl with
long, straight hair and a broad smile and a mortarboard
on her head, clutching a scroll. Sophie holds the cutting
up against the mirror and looks at herself, and then at the

photograph, then back at herself. Ten years separate the girl in the picture and the woman in the mirror.

> Local girl Sophie Wickham has graduated with honours from a top university just three years after the death of her parents in a horrific car crash that almost claimed Sophie's life too.

'You don't look that bad a person to me,' she murmurs, not even knowing why she says that. Then she carefully puts the clipping back within the pages of the book, as though it is some secret thing she must hide away, only for herself to look at.

–

After lunch, Sophie goes up to the Post Office to see Carol, who is stacking up the copies of that day's unsold Sunday newspapers. She breaks into a broad smile when she sees Sophie. 'Hello, lovely. I was just about to close up for the day. Want to come in for a cup of tea? I've got a nice bit of red velvet cake, too.'

In Carol's flat above the Post Office, Sophie nibbles the cake and says, 'Carol, what happened to Margaret? Did anyone ever find her?'

'Thaddeus said he was going to tell you that story,' nods Carol. 'Didn't he say? Nobody heard from her again.'

'But didn't anyone try to find her?' presses Sophie.

Carol shrugs. 'It had never happened before, and it hasn't happened since. We don't know what became of Margaret. She never came back to Withered Hill, and nobody from Withered Hill ever found her.'

'But people tried? To find her? And bring her back?'

'It was a long time ago,' says Carol. 'I wasn't even born. Nothing bad came of it, as far as we know. More tea, love?'

Sophie waits until Carol has refilled her cup, then says, 'And after that…? Someone else came? A new girl?'

'Aye. That's the way of things in Withered Hill.'

'Was Sophie Wickham a bad person?' says Sophie suddenly. She is beginning to feel that outside life inside her, growing, filling her. With that comes what she can only think of as memories, things that had been done. Terrible things in some cases. Secret things.

Carol laughs lightly. 'That's a funny thing to say.'

'Why?'

Carol looks at her seriously. 'Because you're Sophie Wickham.' She takes a sip of tea, her eyes never leaving Sophie's. 'Perhaps that's something you need to think hard and long about, love. I mean, how can you ever hope to leave Withered Hill until you reconcile the fact that the girl whose life you read about on all those papers and printouts is not a different person from you?'

Sophie thinks about it. 'So if Sophie Wickham is a bad person, and I'm Sophie Wickham… that means I'm a bad person too?'

'Do you think you're a bad person?'

'No,' says Sophie, then after a moment's consideration, 'Maybe *she* was. Maybe I *was*.'

'Well, then,' says Carol. 'Maybe we're getting some-where, then.'

'I don't understand.'

'Think on it a bit.'

'Carol, what's a bower?'

'I can't tell you, love. But I think you're getting closer. More cake?'

Back at her cottage, Sophie gets out all her files, all the cuttings and documents and printouts, and lays them out on the rug in the living room. A patchwork picture of a life. A scattering of jigsaw pieces. An out-of-focus blur, slowly resolving into a sharp, clear image. And added to all that... her memories. Her dreams. Snatches of the other life that spring, unbidden, into her mind. Being told her baby sister was dead and not feeling as sad as she should. Telling lies that ruined a girl's life. Hateful, evil thoughts. She has started writing them down as they arrive, because often they disappear again, popping like bubbles on the surface of a pond. She has the overwhelming sense that Sophie before, Sophie Outside, as she has started to think of her, was not a good person. But has she, Sophie now, Sophie Inside, been any better? It's all rather a matter of perspective, she thinks. It depends on who is setting the rules and boundaries.

Sophie finds a big piece of blank paper and with a pen divides it into two columns. At the top, she writes BAD THINGS. Over column one, she writes OUTSIDE. On the second column, INSIDE. Then she sets to work.

BAD THINGS

OUTSIDE	INSIDE
Drinks too much	
Shallow	
Treats people badly	
Self-absorbed	
Wasteful - clothes, etc	
Liar	
Cheat	

Sophie pauses with her pen over the INSIDE column. Is she any better as a person? All she can remember, really remember with any clarity and surety that she has actually lived those things, is the last year in Withered Hill. And she only has the frame of reference of Withered Hill's curious morality, which she knows, deep down in her gut, is not the morality of the outside world. She watched Catherine kill Jamie. She did not do anything to stop it. Everyone had cheered. But Catherine said she hadn't committed murder in Withered Hill, she had made sacrifice. According to the traditions and laws of the place. But Sophie had done nothing. She had watched.

'Doing your homework?'

Sophie jumps. She hadn't heard Catherine let herself into the cottage. Catherine leans over her shoulder to look at what Sophie had written.

'There's still a lot of… duality,' says Catherine. 'Sophie Inside and Sophie Outside. There's still reconciliation to do before you're ready.'

'What is there to reconcile? Sophie did bad things outside. Sophie does bad things inside.'

'Sophie should stop referring to herself in the third person.'

'Can you not be a teacher for a minute?' says Sophie angrily. 'I'm not in your classroom now!'

'Ah, but you are,' says Catherine. 'Isn't that the entire point?'

'So I'm here to learn? But to learn what?'

Catherine puts out a hand to Sophie's cheek. Sophie pulls away sharply. Catherine says softly, 'You didn't pull away from me at Faunalia.'

That was four months ago. Sophie cannot forget Catherine pushing her up against that tree, dragging up the cashmere dress, and sinking to her knees in front of her. Then running, hand in hand, drunk and heady, back to Catherine's cottage. It is the first time they have spoken of it.

'There are some places in the world where that would be a sin, too,' says Sophie. 'Some places it would be a crime.'

Catherine smiles wryly. 'You *have* been doing your homework.'

'The point is,' says Sophie, exasperated, 'it's all a matter of perspective, isn't it? What happens in Withered Hill might be wrong to the outside world. What happens out there might be wrong to us. Who sets the rules?'

'Not people like us. We just play the game.'

Sophie bends down and scoops up a fistful of papers. She brandishes them at Catherine. 'All this is just... It's just meaningless, isn't it? It's done. It's happened. I'm not that person. These things might have happened, might have been done, but I'm not that person. I'm me, Sophie Wickham, right now. I'm not the sum of the things I've done. I'm not defined by the things I've yet to do.' She flings the papers in the air and they rain down like over-sized confetti. 'I wish I could just make all this disappear. All these memories of things I don't really remember. I wish I could just lock them away somewhere and walk out of Withered Hill and be me, me as I am right now, me as I want to be.'

Catherine is smiling at her, which Sophie finds both encouraging and annoying.

'What?' she snaps.

'I think you're getting there,' says Catherine. She moves closer to her. Everything feels suddenly magnified. The sun lancing in through the windows, the dust motes dancing in the rays. The birdsong outside. A dog barking somewhere, a van rumbling on Main Street. The scent of Catherine, woody and earthy and sweet. 'What is it you want to be? Right now, what do you want?'

'I want you to be pushing me against that tree,' says Sophie, her voice little more than a whisper. 'I want you to be pulling up my dress. I want to feel like that again. I want to feel brand new.'

Catherine is so close, Sophie can feel her breath on her cheek. She says, 'You are brand new. Ever since you walked into Withered Hill. But you still carry those things, those memories, those actions. And you can't walk out until you can do so without them.'

Sophie feels Catherine's body against hers, feels the heat emanating from her. She puts her hands on Catherine's hips, pulls her closer, tighter. She murmurs in her ear, 'Do you do this with all of them? All of them like me?' Catherine is unbuttoning Sophie's shirt. 'Is this part of the game?'

'No,' says Catherine, planting a tiny kiss on her ear, then another. 'This is very much breaking the rules.'

Sophie gasps as Catherine's fingers find her bare skin. 'But I thought… I was free to do as I pleased, in Withered Hill. With anyone.'

'Not Withered Hill's rules,' says Catherine, nuzzling Sophie's neck and raising goosebumps on her flesh. 'My rules. I don't get involved with people like you.'

'Why?' says Sophie, closing her eyes.

'Because when you're ready, I'll be the one to take you away from Withered Hill. And I'll never see you again.'

26

Outside

Days to Withered Hill: 7

When Tom Gisburn walked into Sophie's flat, she held up her hand. She said, 'I know this will look mad when you see it. I hope it is mad. But I need you to look at something for me.'

'Okaaaay,' said Tom, his brow crinkling.

Sophie stepped to one side so that Tom could see the pentagram carved into the floorboards. He stared at it, seemingly uncomprehendingly for a moment, then said, 'Wow. Bit of interior decorating?'

'Take a closer look,' said Sophie, and Tom stepped forward and squatted down at the edge of the circle, his fingers tracing the nearest symbols.

He looked up at her. 'Did you do this?'

'No. Did you?'

Tom laughed, then frowned. He stood up and faced her. 'Sophie, are you being serious? You think I did this? When?'

She shrugged. 'Maybe the other morning. When you stayed over. Before I woke up. I found it then.'

'And you think I did it,' he said flatly. 'For what reason?'

She suddenly felt very stupid. 'I don't know. It's just… you're the only person who's been in my flat for months. I just… I just had to be sure.'

'And it wasn't there the day before?'

'I don't know,' said Sophie wretchedly.

'When was the last time you looked under the rug?'

Sophie felt like a small child, being asked to explain something preposterous to a grown-up. 'I'm not sure. Not recently. The rug was here when I moved in.' She took a deep, ragged breath. 'I might never have looked under it, if I'm being completely honest.'

Tom exhaled slowly. 'OK. For the avoidance of any doubt, I did not do this, and I never would do anything like this.' He stared at the pentagram. 'What even is it? Some kind of devil-worshipping shit?'

'Maybe. I don't know.'

His eyes met and held hers. 'I think maybe I should go.'

'Oh, God,' breathed Sophie. 'I'm sorry. I'm so sorry. I know you didn't do it. It sounds ridiculous now. But when I found it… it freaked me out. I put two and two together and got… I don't know. A bazillion.' She looked at him and bit her lip. 'I'm nuts. That must be what you think. You can go if you like. I wouldn't blame you.'

'I have a dinner reservation for eight,' said Tom. 'And you have an absolutely killer little black dress. It would be a shame to waste them both, wouldn't it?'

–

The restaurant was dark and posh and eye-wateringly expensive and Sophie didn't understand what half of the dishes were so just ordered something she was fairly sure had chicken in it. It had an underground car park, too, and Tom had left his car there. He ordered a bottle of wine that cost about three days' pay for Sophie, and said, 'I'll leave the car here and get an Uber home.'

Their eyes locked for a long moment. Sophie didn't say he could stay over, not yet. But she knew that she would. She knew that Tom hadn't done that stupid pentagram. She knew that she wanted, already, to feel the weight of his body on hers.

'You know,' he said, when the wine came. 'Maybe we should call the police if you really think someone's been in the flat?'

The *we* made her feel slightly flustered and warm and nice. Were they a *we*? Already? On their second date?

She sipped the cold wine. 'What would I say? The same I said to you? That I've no real idea how long that thing has been there. Then the landlord would get involved and...' She rolled her eyes. 'I'd probably get evicted.'

'You told anyone else about it?'

Sophie shook her head, then remembered Colin. 'Oh, yeah. This guy at work. Colin. He's a little odd. But he seems to know about this occult stuff. I sent him some photos.'

'And?'

Sophie shrugged. 'It was only last night. He said he'd look into it. He said he thought it might be something used for protection.'

'Protection against what?'

'Exactly. No idea. It's crazy, isn't it?'

Tom pulled a face. 'It's weird, that's for sure. What exactly is it you do at that place, anyway?'

'Data entry. Odd stuff. We get bundles of typed, some-times handwritten, information. Numbers, sometimes, or words. A lot of it meaningless. Sometimes in foreign languages. I have to input it.'

'Clever,' said Tom, thoughtfully.

Sophie laughed. 'Not really. A monkey could do it.'

'And how long have you worked there?' asked Tom, just as their starters arrived.

Sophie poked around in her crab linguine. 'Not even a month. I was temping for a while. I had another job I liked, at an insurance office, but they had to get rid of loads of people.'

'And how did you find this place?'

Sophie frowned. 'It was a bit weird, actually. I was out with my friends. Colin approached me outside a bar. Said he'd overheard me talking about needing a new job. He gave me the card of his boss. I called her, got an interview the next day, and started the day after that.'

'Lucky break,' said Tom.

Sophie nodded. 'I suppose it was.'

Tom finished his starter and laid his knife and fork on the empty plate. 'You know, if you don't like it there, maybe I could find you a job with me?'

'I didn't say I didn't like it,' replied Sophie, as Tom poured her another glass of wine. 'Besides, it's not really a good idea to mix business and pleasure, is it?' She put her hand across the table and stroked his wrist lightly. He gave her a long, piercing look that meant both of them knew he wasn't going anywhere tonight except Sophie's flat.

'I'll fix us some drinks,' said Sophie as she turned on the lamps in the living room. 'Wine OK? Gin?'

'Wine is fine,' said Tom. He was staring at the penta-gram. Sophie wished she'd covered it up with the rug before they'd gone out. It made her feel stupid for accusing him all over again.

'Make yourself comfortable,' she said. 'I'll be one sec.'

In the kitchen, she kicked off her heels. She was feeling a little drunk. More than a little drunk. In the cab on the way back, she'd taken Tom's hand and held it on her lap. She felt like she was on fire.

When she came out of the kitchen, Tom was standing by the sideboard, looking with some bemusement at the little stapled book that held all her workplace contacts. He said, 'Very old-school. Have they never heard of databases at your place? Bit odd for a data company.'

Sophie handed him the glass and took the book from him, dropping it on the sideboard. 'Can we not talk about work on Saturday night, please? Unless it's about your latest covert mission to somewhere glamorous to save the world from a despotic billionaire baddie.'

He moved closer to her, and she smelled his woody cologne. 'I'm not a spy, I'm afraid. It's far more boring than that. Would you like me more if I was, though?'

'Oh, I like you well enough as it is,' said Sophie. She pulled away from him and walked barefoot across the pentagram, feeling the scratched symbols beneath her feet. She stood in the middle and turned to him. 'I slept in the middle of this last night, you know.'

'Now that is weird,' said Tom. 'Given that you've got an entirely serviceable and, indeed, quite comfortable bed.' He raised one eyebrow, which drove Sophie slightly wild. 'Speaking of which...'

She put her glass down on the coffee table, which she'd pushed against the wall, by the rolled-up rug. 'Oh, I think I'm fine here. This being a circle of protection, and all. Maybe it's to protect me from you.'

He smiled and his eyes danced. 'Yes, maybe you're right. To be honest, I think you need protecting from me right now.'

'Is that a fact, Mr Gisburn? Are your intentions not entirely honourable?'

He laid his own glass down on the sideboard. 'No, Ms Wickham, I'm very much afraid that they aren't.'

'Well,' she said. 'In that case, perhaps you should come here and ravish me where I stand.'

He pulled a sad face and spread his arms wide. 'I'd like nothing more but… your circle of protection and all that. I very much doubt I'd be able to cross it unless you invited me.'

Sophie giggled. 'That's Dracula, isn't it? Anyway. I invite you across my threshold. Into my magic circle. I invite you to come and fuck me.'

Tom smiled and stepped into the circle, walking over the symbols and carvings, and stood in front of her. He looked her up and down from the top of her head to her feet, and back again. Then he took the shoulders of her dress in his hands.

'If I were to promise to buy you a replacement,' he murmured, 'would you be at all amenable to me ripping this dress from your body?'

Sophie shivered and felt her stomach flip.

'Do it,' she breathed.

He did.

Days to Withered Hill: 6

Sophie woke to the smell of coffee and bacon, and Tom, fully dressed, sitting on the edge of her bed with a tray.

'That's real coffee,' she said. 'And bacon sandwiches.' She frowned. 'I don't have any real coffee. Or bacon.'

'I went out to the shop while you were sleeping. I saw that old boy from across the hall. Got him a newspaper and some milk. He's a nice guy. You should watch out for him more.'

She sat up and rubbed her eyes, pulling the duvet up to hide her naked chest. 'God, I must have been dead to the world.'

'It was a late night,' said Tom with a smile. Sophie could attest to that, as could the tenderness between her legs. A very late night.

'You aren't eating?' she mumbled through a mouthful of bacon as she wolfed the sandwich down.

'I had one in the kitchen,' said Tom. He glanced at his watch.

'You have to go,' she murmured. She'd had some fanciful notion of them spending the day together. Maybe going for a walk in the park. Some romantic bullshit like that.

He nodded. 'I have work to do.'

Sophie took a deep breath. 'Tom. Please tell me now before this goes any further. Are you seeing anyone? Or, worse, married?'

He leaned forward and wiped a spot of ketchup from her chin with his thumb. 'No. I promise you. But I do have a very demanding business that does not keep to office hours.'

'Married to the job,' said Sophie, smiling.

'In a way.' He glanced at his watch again. 'Look, how about we do something... Wednesday night? And... if you wanted to, if we both wanted to... maybe you could stay over at mine? I could drive you into work the next day.'

Sophie smiled with a relief and gratitude she found almost unbecomingly desperate. She composed herself and said, 'Yes, that would be lovely.'

'It's a date, then,' said Tom. 'I'll call you tonight, if you like? If you haven't heard enough from me?'

'That, too, would be lovely,' said Sophie. In fact, everything was turning out rather lovely all over.

27

Outside

Days to Withered Hill: 5

They'd talked on the phone all night, or at least it felt like it. Sophie awoke for work with a wine headache and a nerve-jangling lack of rest. She had drunk too much and slept too little and couldn't remember half of their conversations, but woke up with a warm, tingling sense that this time, with this man, things might actually be OK.

She wasn't sure if she could wait until Wednesday to see Tom again, but she wasn't going to push him for an earlier date. *Play it cool*, she told herself sternly. *Go with the flow. Do not – I repeat do NOT – fuck this up.*

She got into work showered, refreshed and on time, fuelling her recovery with a big cup of coffee. She'd just opened up the folders laid on her desk to look at the day's work when Mandy popped her head around the door of her office and called for her to come over.

'Before we start,' said Mandy, pushing a piece of paper across the desk at her, 'I need you to sign this.'

Sophie glanced at the form and her eyes widened. 'The Official Secrets Act? Is this a joke?'

Mandy stared at her. 'And how often have you known me to make jokes?'

'But...' Sophie looked at the paper again. 'The Official Secrets Act?'

'We have a wide range of clients we do work for. Most of the time, it's for the scientific or business communities. Occasionally... it's for someone else.'

'The government?'

Mandy sighed. 'I literally cannot say another word unless you sign that form. If you're unwilling, then I will find someone else to do the job. I should point out that it comes with a rather hefty bonus.'

Sophie bit her lip. It all sounded very mysterious. But then, think of how much fun she'd have telling Tom. Oh... she wouldn't be able to tell him, would she? What if he actually was a spy? And if she told him about this, then he'd know and get her into trouble and...

'Sophie?' said Mandy.

Sophie blinked. 'Sorry, yes.' She took the pen Mandy was offering and scrawled her name at the bottom of the form and pushed it back across the desk.

'Excellent,' said Mandy, putting the paper in her desk drawer. 'Now. We have a very special job come in, last minute as these things often do. It is, as you surmise, a government job. Intelligence services.'

'Wow,' said Sophie.

'Very, very sensitive information. And, as a result of that, we can't process it here.'

Sophie frowned. 'So, where?'

'A secure facility. In Cornwall. Very remote.'

'Cornwall?' said Sophie. 'That's a long commute.'

'You have to stay in residence until the job is completed. It shouldn't take longer than a week.'

Sophie was having trouble processing all this. 'A week? I have to go and stay on... what, some army base or something? For a week? On my own? Why have they come to us with this? Don't they have their own data processors?'

'It's a long-standing contract,' said Mandy. 'No, it's not an army base. It's a house. A big, old house miles from anywhere. And you won't be alone. I'm sending Colin as well.'

Sophie pulled a face. A week in Cornwall with Colin. 'When is it?'

'You leave on Wednesday.'

Sophie's face fell. 'The day after tomorrow? But I have a date on Wednesday.'

Mandy shrugged again. 'Take it or leave it. I mentioned a bonus. It'll basically be about two months' salary for you.'

Sophie opened her mouth and shut it again. An extra two months' salary. And it was only a week. Surely Tom would understand.

'You can't tell anyone where you're going, or why,' said Mandy. 'Not even your date. You'll have to come up with some excuse. You've just signed the Official Secrets Act, remember.'

Sophie thought about it, then said, 'OK. Yes. I'm in. What do I need to do?'

Mandy smiled. 'I'll organise train tickets today. In the meantime, you have work on your desk. You can take tomorrow off to pack. I'll call you with the train times. Can you send Colin in when you go out?'

'He wasn't at his desk,' said Sophie.

Mandy pulled a face. 'Not like him to be late. I'll call him.'

It was almost lunchtime and Colin hadn't yet showed up when two people were escorted into the office by the receptionist: a woman in a dark suit with tied-back red hair and a tall man in a raincoat, carrying a briefcase. Sophie watched as they were taken into Mandy's office and sat at her desk, talking with her for a good half-hour. Then Mandy came to her door and waved at Sophie, beckoning her over.

'Are you all right?' said Sophie when she got to the door, seeing Mandy's stricken face. 'Has something happened?'

'These people need to talk to you,' said Mandy. 'It's the police.'

Sophie felt the colour drain from her face. 'The police? What's happened?'

'Just talk to them,' said Mandy. She dropped her voice low. 'Don't mention Cornwall. Official Secrets. That's higher than the police. It's nothing to do with that anyway.'

Mandy left the office to let her in and closed the door behind her, leaving Sophie with the two people now sitting on Mandy's side of the desk. They indicated for her to sit down.

'My name is Detective Chief Inspector Parish,' said the woman, her mouth a thin, bloodless line. 'This is Detective Inspector Coonan. You are Sophie Wickham, correct?'

Sophie nodded wordlessly.

The woman looked at her sharply. 'Correct?'

'Yes,' said Sophie. She was suddenly burning with unexplained guilt. 'Have I done something wrong?'

She thought, suddenly and unexpectedly, of when Emily had died. That strange policewoman. Sophie had never been sure that exchange between them had really happened.

'*Sophie? Do you know what happens to bad girls?*

'*Sometimes. Sometimes they go to prison. Sometimes, very bad girls who continue to be bad, though… well, something else happens to them.*

'*Sometimes, Owd Hob takes them for his wife.*'

'You work with Colin Turpin, right?' said the man, Coonan.

'Yes,' replied Sophie, blinking, pushing away the memory, and realising she hadn't even known Colin's surname. 'Yes, I do.'

'Ever fraternise with him out of work?' asked Parish.

Sophie shook her head. 'No. I don't know him very well at all, I'm afraid. Has he done something wrong?'

Coonan opened his briefcase and handed a sheaf of papers to Parish, who looked at them, then put them on the desk in front of Sophie. 'These are text messages you exchanged with Mr Turpin on Friday evening, correct?'

Sophie looked at the printouts. They were her Whats-App messages to Colin. She looked up at Parish. 'What is this about?'

'Correct?'

'Yes, correct. Those are my messages.'

Parish nodded and Coonan handed over another sheaf of paper. These were the photos she'd sent to Colin of the pentagram. Coonan said, 'You asked him to look into this for you? You had found it on your floorboards under a carpet?'

'A rug. Yes, I asked him.'

'Why?' said Parish.

'I don't know. He seems to know about this sort of stuff.'

'I thought you said you didn't know him very well at all?'

211

'I don't,' said Sophie. 'Look, are you going to tell me what this is about?'

Coonan took out another sheet of paper. This time, it was a photograph of two pages from what looked like an old book, yellowed and crinkled at the edges. There were columns of what appeared to be handwritten words, barely legible. And not even English. Latin, Sophie guessed. But the pages were dominated by a large diagram of what was unmistakably the pentagram on Sophie's floor.

'Oh,' she said. 'So he did find something out.'

'As far as we can determine, this book is very rare,' said Parish. 'And it has been in Mr Turpin's possession for quite some time.'

Parish looked at her for a long time, saying nothing. Sophie was starting to feel very uncomfortable. Eventually, Coonan showed her another photograph. This time it was of Sophie, outside her flat, putting rubbish in the wheelie bins. She recognised it as from Saturday morning. It was taken from across the road, blurred traffic whizzing past in the foreground. She remembered that sense of seeing someone from the corner of her eye, someone she recognised.

'Where did you get this?' she asked.

Coonan showed her several more photographs. Sophie sitting on the bench drinking coffee and eating lunch. Sophie walking towards the Green Man to meet Tom Gisburn. Sophie running in the park. Sophie sitting at her desk. The penny finally dropped.

'Colin took these? All of these?'

Parish nodded and said, 'Do you have your house keys on you?'

Sophie pulled them out of her jacket pocket and handed them over. Coonan took from his briefcase a clear plastic sealed bag, in which was a single key, attached to one of those labels they used to put on luggage. Written on the label was her name. Coonan held up the key from her fob against the one in the bag and squinted at them. 'Looks like a perfect match.'

Sophie gaped at him. 'Colin has a key to my flat?'

'How do you think that could have happened?' said Parish. 'Did you give him one?'

'No!' said Sophie. 'I've no idea how... oh.'

Then she remembered. Colin spilling the coffee on her desk. Then she'd seen him at the key-cutting booth. Then he'd kicked her bag over. He'd taken the keys the first time, had a copy made, and given them back to her while pretending to pick up her stuff.

'The sneaky little bastard,' she said. She looked at Parish. 'He's been in my flat, hasn't he? He's the one who scratched that bloody pentagram in my floorboards.' She nodded furiously to herself. 'He knew I was going out on Thursday night. He did it then.'

Coonan said, 'It also appears he set up a number of social media accounts with the express purpose of targeting you. Have you had any communications on any platform from an account using the name...' He consulted a notebook. 'coldiron6239745, or some variation thereof?'

Sophie felt anger rising in her like the mercury in a thermometer. 'Where is he? Have you arrested him? I want to press charges. Whatever they are. I want him locked up.'

Parish's face seemed to soften a little, and she glanced at Coonan. She said, 'Sophie, Colin Turpin has been found dead.'

Sophie's hand flew to her mouth. 'Dead? How? What happened?'

'He was discovered in his flat yesterday evening by a neighbour who was concerned she hadn't seen him for most of the day,' said Coonan. 'It appears he might have inadvertently taken his own life.'

Sophie frowned at him. 'What?'

Parish said, 'I'm sorry, Sophie. This must be very upsetting for you. His body was found in a state of undress with several photographs of you scattered about. He had a plastic bag over his head. He appears to have been in the process of an act of extreme autoeroticism that went terribly wrong.'

Sophie screwed her face up. 'Oh my God.'

'There'll be an inquest, but we'll try to make it so you don't have to give evidence. We will need to take a formal statement, though. We'll be in touch about that.'

Coonan gathered up the papers and snapped shut his briefcase and the two stood up.

Parish said, 'I'm really sorry about this, Sophie. There are some fucking weirdos about. But there's no evidence he would have actually caused you any harm. And even if there was...' She shrugged. 'He can't hurt you now, can he?'

The two police officers left, and Sophie watched through the glass partitions as they spoke briefly to Mandy and then were taken downstairs by the receptionist. Mandy came into the office and closed the door behind her, and exhaled loudly.

'Bloody hell,' she said. 'Are you all right?'

Sophie nodded numbly. 'It's a lot to take in.'

Mandy sat heavily at her desk. 'I mean, I always knew Colin was a little... unusual. But this...'

Sophie nodded again. 'So I suppose Cornwall is off now?'

Mandy shook her head. 'This isn't the sort of job that can be put off. And we can't afford to lose it. I'll understand if you don't want to do it now, though. I'll get someone else.'

'No,' said Sophie firmly. 'I'll do it. It'll take my mind off the fact that creepy fucker has been in my flat. Will I have to go on my own?'

'No, I'll come with you,' said Mandy. 'I don't want to really trust this to anyone else at this late stage.' She looked at her phone. 'Look, take the afternoon off. Get ready for tomorrow. I'll email you the train details later.'

Sophie nodded and turned to leave, and Mandy said, 'And remember. Tell no one where you're going.'

28

Inside

Days in Withered Hill: 138

'So what happens at Halloween?' says Sophie. She fancies there will be, like she has heard tell of all of Withered Hill's many festivals, bonfires and revelry. And the perfect opportunity to try to escape the village.

'Not Halloween,' says Zachary Winterbottom, rolling his chewing tobacco around his mouth and letting loose a stream of it through stained teeth. 'Samhain.' He says it *Sawin*, though that's not what it reads like when Sophie looks it up later in the library. 'Halloween is plastic and fun and pumpkins and silliness. Samhain is… something else.'

They are walking along the muddy road past Nut Nan Farm, where Sophie can see Peter O'Keeffe and his dog wrangling sheep into a pen in the top field. Zachary hauls himself along with a long, gnarled stick, worn smooth by his grip over many years, many decades probably. Sophie has on her wellies, and a scarf around her neck, though it's mild and clear.

'Not right, being so warm at the end of October,' mutters Zachary. 'I've known it to snow on Samhain, when I were a lad.'

'I'm not complaining.'

'Aye, well, you should. This is what comes of the outside's disregard for the natural order of things. Of its meddling and greed and selfishness.'

The old man had called unexpectedly at Sophie's cottage and asked her if she wanted to walk with him. 'Where to?' she'd asked.

Zachary had shrugged. 'I don't know yet. Does it matter?'

He leads her to the edge of the woods – on the road where she had tried to hide out with Peter's sheep in one of her disastrous early attempts to leave – and then turns left, walking along a rough path between the treeline and Nut Nan Farm's low fences. He glances sidelong at her and says, 'Tha'll be trying to leave again tonight, no doubt.'

Sophie says nothing. A few days ago, she watched a film on TV, *The Great Escape*. Allied military men in a prisoner-of-war camp. They saw it as their duty to escape. Sophie feels the same, to some extent. She knows her place isn't here, in Withered Hill, but out there. And while the four-and-a-bit months have not been onerous – in fact, she is rather enjoying her life – she is driven by a compulsion to try to get out.

'Does everyone try to escape? Everyone like me?'

Zachary shrugs as he walks. 'Mostly. At first. Then they knuckle down and concentrate on what's important.'

'And what is important?'

He stops, leans on his stick, and looks at her with a frown. 'Why, working out *how* to leave.'

Sophie sighs. 'You're talking in riddles.'

'No, I'm not,' he says crossly. 'You're *listening* in riddles. There's a difference between escaping and leaving, you know. The door is wide open for you, Sophie Wickham. You just have to work out how to cross the threshold.'

They circle around Nut Nan Farm, the ring of trees on their right. Sophie peers into the dark, thick woods, sometimes catching a movement on the periphery of her vision. Zachary watches her and says, 'Aye, they're active today. Samhain. That's why it's not a good idea to try to leave tonight.'

'Are there festivities?' asks Sophie.

He shakes his head. 'Not of that sort. But it is a very important night. For everyone in Withered Hill.' He looks her dead in the eye. 'Everyone in Withered Hill except you. Take my advice, girl, you'll lock your doors and windows tight and have an early night, and try to ignore anything you hear or see outside your little cottage. Now be off with you. I have business.'

Zachary walks over to where a spade is leaning against one of the trees. Sophie says, 'What's that for?'

'Digging holes,' says Zachary, looking at her as though she's stupid.

'And you're digging a hole in the woods?'

Zachary nods. 'Aye. Want to get it finished before the ground gets too hard and frosty. Needs to be ready for Faunalia.'

'When is Faunalia? And what is it?'

'February. And you'll see soon enough.'

Sophie looks at the spade in Zachary's thin, bony hands. 'Perhaps I could help you?'

He shakes his head. 'No. Thank you, but this is one hole I have to dig myself.'

—

Sophie spends the afternoon searching her growing social media files for mentions of life outside at Halloween. There are photos – plenty of photos. Sophie dressed as a witch, with long, artfully ragged dress and plunging neckline. Sophie as a vampire, distended canines, wearing a black, lace-up bodice. Sophie as a zombie schoolgirl, blood on her chin, ripped white blouse, tartan miniskirt. When Jamie arrives on the scene, Sophie's costumes become a little more demure, a little less revealing. There are photos of them at a party, in boiler suits, with backpacks and ghost-hunting ray guns attached. Another time, they are both Egyptian mummies, wrapped head to toe in bandages. In the earlier photos, Sophie looks free and abandoned in her revelries. With Jamie, there is a more guarded look in her eyes, a sense of endurance rather than enjoyment. She likes the joyful Sophie a lot more.

Earlier, Peter O'Keeffe had been round with some lamb chops. Sophie never carries money in Withered Hill; she goes into one of the village shops and is given what she needs or asks for. Catherine turns up with gifts and clothing. Food fills her fridge and cupboards. It is as though Sophie is a child of Withered Hill, a shared responsibility, to be looked after and cared for by the community, even as they keep her prisoner in the confines of the ring of dark, thick woods.

Peter flirts with her every time he visits. He is clumsy and none too subtle, and he undresses her with his eyes, without any shame. He is married, or handfasted, at least; they do not marry in Withered Hill, or not in the conventional sense that Sophie understands. The church is now a library. The school where Catherine teaches is called St Michael's, and that is a nod to the Christian faith, but only because, as Thaddeus Obermann once told her, Michael was the patron saint of fairies and intervened with God on their behalf, and saved them from destruction so long as they inhabited dark, out-of-the-way places.

Sophie stares out of her window as dusk falls at the woods, dark and out of the way. She thinks of Peter as she cooks her dinner, thinks of his strong arms, his broad, farmer's back. Wonders what she would do if, on one of his visits, he scooped her up in those arms and carried her off to bed.

As soon as she has sat down at the kitchen table with her meal, Catherine lets herself in, carrying a cardboard box. She dumps it on the sofa and peers over Sophie's shoulder at her dinner, pinching a thick-cut chip. 'Mmmm, good,' she says. 'I'm famished.'

'Want some? There's plenty.'

Catherine shakes her head. 'I can't stop. Samhain tonight. I'll be eating later. Brought you something.'

Sophie leaves her meal and goes with Catherine to the cardboard box, which contains a dozen hollowed-out turnips with grotesquely carved faces. Inside each one is a tealight.

'Light them and put one in each of your windows and on each doorstep, front and back,' she instructs.

'Is this a Samhain thing?' asks Sophie. 'So I am involved?'

Catherine smiles. 'No, this is to ensure you aren't involved. Samhain isn't for you.'

Sophie pouts. 'Why not?'

'It just isn't,' says Catherine. 'The turnips will make sure you aren't visited.'

'Visited by who?'

'Just light them and do as I say,' instructs Catherine. 'And don't go out tonight. Just go to bed early. And if you hear anything, just ignore it.'

Sophie pouts. 'Is this because of Lammas?'

Catherine frowns. 'What do you mean?'

It has been three months since the Lammas festival, and they haven't really spoken of it. After Sophie ran, appalled and terrified from the square, she had locked herself in her cottage and refused to come out for three days, until she had run out of food. When she did emerge and she eventually saw Catherine again, she had cowered away from her.

'What's up with you?' Catherine had said.

'That bread... the meat in it...'

'Hares. I told you the farmers had been rounding them up for days.'

'It was disgusting.'

Catherine had shrugged. 'It was Lammas.'

And that had been an end to it, until now.

Sophie says, 'Is this a punishment because I freaked out at Lammas? Not letting me take part in Samhain?'

Catherine laughs. 'You are funny.'

That makes Sophie more angry. 'It wasn't funny. You were all covered in blood. From eating those hares in the bread. But that wasn't the worst part. The worst part was the look on your face. On all your faces. You looked... savage.'

Catherine puts a hand out and touches Sophie's cheek, and it is all she can do not to flinch. 'We are savage, Sophie. All of us. We dress ourselves up in nice clothes and read interesting books and talk about the state of the world, but deep down we're all savage. It's just that in Withered Hill we haven't forgotten that, or try to pretend otherwise.'

She puts her head on one side and looks at Sophie for a moment. 'The world is green and red, Sophie. Green and red. That which grows from the earth, and that which walks upon it. The two are inextricably linked, like...'

'Like a hare baked into a loaf?'

Catherine nods and smiles. 'Exactly that. Like a hare baked into a loaf. Outside, they've forgotten that, mostly. They eat meat, but they don't want to know where it comes from. They don't want to look at the steak on their plate and think of the cow in the field, eating grass. You can only be one with the natural world if you accept that it is all part of the same cycle.'

'And Samhain? Is that the same?'

'No,' says Catherine with a tight smile. 'Samhain is not the same. Samhain is about remembering that those we have lost never really go away forever.'

29

Outside

Days to Withered Hill: 5

When Mandy let Sophie go early from the office to prepare for Cornwall, she hit the shops to get some toiletries and a couple of books to while away the time. She wasn't expecting Mandy to be scintillating company. Mandy had told her to finish one last job before she left, a page of two-syllable Spanish words that simply needed typing in. Sophie had stared at the sheet for a moment, then took a photo of it on her phone and stuck the page in the tray to say it was done. She'd do it from home later. She was oddly excited about the trip away and wanted to be off and packing. As weird as the whole thing was, it was the nearest she'd get to a holiday this year.

As she perused the make-up counter, she wondered if it would literally be her and Mandy, or if there would be other people there as well. She had no idea how big the company was, whether their office was just a tiny cog in a huge machine. A cog with one less component now.

Sophie didn't quite know how to feel about Colin. On the one hand, she was glad the slimy little creep was out of her life. On the other... he'd died. Killed himself. She tried on a shade of red lipstick and pursed her lips in the little mirror. She should feel… well, something about that, surely.

Her therapist, when she could afford to visit her, had once told Sophie that she had an unusual relationship with death. That she seemed capable of blocking it out, pushing it away. The therapist had wondered if this was some kind of denial about her own mortality. No, Sophie had said. I just don't like to think about death. Who does?

She knew, though, that perhaps she wasn't quite normal.

She paid for the lipstick and went to buy some shampoo, conditioner and body wash.

She didn't remember ever being properly upset when her parents had died. God knows her therapist had mined that particular seam for hours on end.

'I was hurt in the crash myself,' Sophie had said. 'By the time I came to and was given the news… I was in shock, really. And once that wore off… well. There was nothing that could be done to bring them back.'

The therapist had gently pointed out that her parents weren't the only family members she had lost.

Sophie headed for the bookshop and perused the latest releases. They were on a three-for-two offer. Sophie made a selection, something not too demanding but hopefully engrossing enough. She wondered what the catering situation was. What if it was miles from anywhere and there was no wine? She felt a momentary stab of panic. She should take some. Maybe a bottle of vodka as well.

As she was paying for the books a flash of purple caught her eye on the edge of her vision. Sophie turned to see the back of a girl in school uniform. It was the same colour as her own high-school uniform. That's why it had grabbed her attention. The girl was maybe twelve, browsing through a pile of young adult novels on a table near the door. She had long, dark hair, the same shade as Sophie's. She turned, slightly, and Sophie was surprised to see that the school tie was the same colours as she had once worn: blue and green stripes. Sophie smiled, taking the bag of books from the cashier. They were a long way from Gloucestershire. It was to be expected, with all the schools in the country, that some of them would have the same uniform.

The girl was staring at her. There was something about her. Something about her eyes, about the curve of her nose. She heaved her rucksack onto her shoulders and moved towards the door, still staring at Sophie. Then she smiled, and raised her eyebrows, and nodded towards the exit.

Sophie glanced around, to see if the girl was looking at someone else, but there was no one behind her. There was something... familiar about her. A sudden ache throbbed in Sophie's temples, and she sensed angular lights at the periphery of her sight. Was this a migraine?

The girl had left and Sophie followed her through the doors, gasping for fresh air.

The schoolgirl was standing opposite, leaning on the wall of a closed-down bank branch with a TO LET sign on it. She was facing Sophie, just staring at her. So close that Sophie could see the badge on her purple blazer.

It was her school's crest.

Sophie frowned at the girl, squinting at the ache in her head. Could it be the daughter of someone she knew from back home? Then where were her parents? Why didn't she come and say something? And why did she look so familiar?

Sophie stared at the girl, and the girl stared back. Now she knew where she recognised those eyes from. They were hers, near as damn it. As was the hair. As was the curve of the nose.

'Emily?' said Sophie, barely a breath, even as she knew it was ridiculous.

The girl giggled and a pair of pensioners walked between them, and when they were gone, so was the girl. Sophie looked around, wildly, and saw a flash of purple further along the street. She set off after her, not really knowing why. She had almost caught up with her when the girl skipped into the road, and when Sophie tried to follow, a black cab sounded its horn and she jumped back, trying to keep her eyes on the purple blazer. Sophie dashed across the road between two buses, just as the girl dipped into an alley between the shops.

Sophie paused at the entrance to the alleyway, peering into its shadowed depths. What was she doing? Why was she following this girl? Her head throbbed and throbbed and throbbed, pushing the thoughts out. She had to know.

Don't be ridiculous, she told herself. *It can't be—*

'Emily!' said Sophie, hoarsely, then tried it again, louder. 'Emily!'

There was a flash of movement at the end of the dark alley, and Sophie slipped between the high brick walls, hurrying past an overflowing dumpster in which something shifted and rustled.

The alley seemed to get narrower as she walked briskly down it, and it ended in a corner, dark, almost pitch black. Sophie pulled out her phone and switched on the torch. High above, the sky seemed to darken from its pale blue to a thicker, treacly shade. Sophie took a deep breath and turned the corner, which ended in a doorway that stank of piss. The girl was there, leaning on the door, waiting for Sophie.

'Who *are* you?' said Sophie.

'Who are *you*?' said the girl.

The more Sophie stared at her, the more she looked like Emily. Like the girl Emily would have become. Would have become if she hadn't died that night when Gran was looking after her.

'Sophie,' she said. 'I'm Sophie. And you…'

The girl smiled. It was like looking in a mirror. 'You know.'

'Emily?' Sophie felt light-headed, dizzy, as though the solid ground beneath her feet was lurching away from her. She swallowed drily. 'But you're dead.'

The girl adopted a serious face. 'And how do you feel about that, Sophie? Would you say you have an unusual relationship with death?'

Sophie opened her mouth to speak, then shut it again. That was exactly what her therapist had said. The exact same words.

The girl – Emily – said, 'Do you think it's a reflection of what you worry about in terms of your own mortality?' The girl fixed Sophie with her gaze. 'Did you even cry when I died? When Mum and Dad died? Do you ever cry, Sophie? Do you ever cry for anyone but yourself?'

'Who are you?' whispered Sophie. She felt rooted to the spot. But the ground was moving, like a ship in a storm-tossed sea, making her stomach flip.

'Who are *you*?' said Emily again. 'Why are they after you? What do they want from you?'

Sophie took a step back. 'Who? Who do you mean? The messages? That was Colin.'

Emily pulled a sad face. 'Poor Colin. Would you say you had an unusual relationship with death, Sophie? Would you say you were happy when you heard about Colin? Would you say you cheered inside and your heart soared? Why weren't you sad, Sophie? What sort of person are you?'

Sophie felt herself cringing, becoming smaller, hugging herself. 'He was stalking me... he was a weirdo.'

'You were horrible to him!' shouted Emily, making Sophie flinch. 'You should be a better person! You *will* be a better person!'

'I'm sorry,' said Sophie, her voice barely a whisper.

'You were glad he died!' yelled Emily. Either Sophie was shrinking into herself or Emily was getting bigger, taller. 'You were glad Mum and Dad died!'

'That's not true,' said Sophie, horrified. Emily seemed to be stretching, growing, her limbs and face distended and grotesque.

'You were happy when I died!' said Emily, her voice a shriek.

Sophie closed her eyes and shook her head. 'No. No, no, no. It's not true. I was very small. I was only nine, I didn't understand—'

'Would you say you had an unusual relationship with death, Sophie?' said Emily, except now it was in the voice of her therapist. 'The way you push things into a box in your head and lock it and throw away the key, do you think that's useful?'

'Please stop saying that,' whimpered Sophie. She opened one eye. Emily didn't even appear human any more. She was long and spindly and crouched over in the doorway, her face stretched, her mouth hanging open like a snake that had unhinged its jaw to swallow its prey.

'Can you be a better person?' said Emily, or whatever it was. 'Can you be better? Or is it too late?'

'I'm sorry,' said Sophie, the words erupting in a strangled sob. 'Emily. I'm sorry. I—'

Suddenly, there was a loud ping, and her phone vibrated in her bag.

Emily said, 'That's your hour up. See you next time. If you can afford the price.'

And then she was gone, and in front of Sophie, cowering in the doorway, was an old man, dressed in a ragged suit, piss staining the front of his trousers. He held his hands out as if to fend Sophie off, his eyes closed, crying in a mewling voice, 'Please, please, leave me alone, I haven't done anything to you.'

Sophie blinked and looked down at the tangle of filthy blankets in the doorway, where the homeless old man obviously slept. 'Oh my God,' she whispered. Her phone was still vibrating. She reached in her bag and pulled it out. It was Mandy.

Cradling the phone between her neck and ear, Sophie pulled a ten-pound note out of her purse and thrust it into the old man's hands, then walked away swiftly along the alley, which no longer seemed as narrow and dark as it had. The sky was blue and pale above her, the street at the end thronged with people.

'Mandy?' said Sophie into the phone.

'You didn't do your work,' said Mandy abruptly. 'That last page. You put it in the out tray but it's not on the system.'

'I was going to do it at home,' confessed Sophie. 'I'm sorry.'

Mandy swore. 'Sophie, the work is… the work is for a reason. Have you spoken to anyone since you left? Has anyone spoken to you? You've told no one where we're going?'

'What?' She risked a glance over her shoulder. There was no one following her. No old man. No Emily. The memory of it was already becoming insubstantial, like a dream. 'No. I've spoken to no one.'

Mandy sighed. 'OK. Well. Please never do that again. I'd like you to come back to the office. Do that last sheet. It's important.'

Sophie was about to protest, but Mandy killed the call. Sophie stared at her phone for a moment, then, with one last look behind her, headed back to the office.

30

Inside

Days in Withered Hill: 138

It begins a little before midnight. Sophie has done as Catherine instructed, and lit all the turnip lanterns, placing them in each window and on the front and back steps. But she has been unable to sleep, intrigued by this night in Withered Hill that, for the first time, is denied her. She sits in her bedroom window, looking out over the dark woods, wondering what is to happen on this Samhain night.

At first, she thinks it is her eyes playing tricks, from staring at the dark trees for so long. Then she thinks it is fireflies, or some other luminescent insects, dancing in the trees. And then, perhaps the village children, engaging in some Samhain game, maybe with turnip lanterns themselves.

But the lights number more than Withered Hill's children, and as they emerge from the edges of the trees, there is more to them. They swing on the ends of rods, held by flittering black shapes. It is them. The tree people. Those who reside in the woods. And they are not alone. Each black figure holds by the hand a shimmering blue form, vaguely human in shape at first, but becoming clearer and more defined as they come closer. People, but like jittery old film footage of people, staticky and fuzzy, led willingly by the dark things.

But where to?

To her house?

Sophie scuttles back from the window, then creeps back towards it, peering around the corner. No, not to her cottage. The strange pairings, their way lit by bobbing lanterns swinging on the end of sticks, peel off as they emerge, and Sophie sees one go to the cottage next door, one to the other side, and to each of the houses around her. To all the houses in Withered Hill, she thinks. Every single one except hers.

Sophie has no real understanding of what the things in the woods are. Only that the villagers venerate them, and respect them, and sometimes fear them, and draw their good fortune from them. But now they are not in the woods. They are out, in the village, with those bright, brisk, shining people that Sophie also does not understand.

Curiosity getting the better of her, she opens the front door a crack and peers around at the stately procession from the woods, the lights disseminating around the village. One pair of black shadow and bright light is passing in front of her cottage, and emboldened she opens the door wider, stepping out to see better. Her foot connects with something and the turnip lantern on the step goes skittering along the path. The two figures ahead of her halt, the black shape sniffing the cold air like an animal. Sophie stands stock-still, and holds her breath, and eventually the thing moves on, the staticky form of its companion moving along with it. When they are out of sight, Sophie hurriedly goes back indoors.

She comes awake, startled, in the chair where she has dozed off, the book she had been reading on the rug at her feet. She rubs her eyes. The book must have fallen and woken her. Sophie yawns and stretches stiffly, then freezes. There's a sound from the kitchen, like something scratching on wood. Cautiously, she creeps to the doorway and peers in. The kitchen is in total darkness. On the windowsill, the turnip lantern stands, the tealight burned out. Sophie remembers what Catherine had said. *The turnips will make sure you aren't visited*. She remembers, too late, kicking over the turnip on the doorstep, and when she looks back to the living room window, there's a thin column of white smoke rising from the guttered candle in the lantern there.

The scratching sounds again, so near it makes her jump. The kitchen door. Something is outside, clawing the door like a cat wanting to be let in. Slowly, Sophie backs away from the kitchen, and jumps at a noise behind her. Whirling round, she sees a dark shape at the window, tapping on the glass.

Sophie runs for the stairs, climbing them two at a time, and rushes into her bedroom, slamming the door shut. She dives onto the bed and wraps the duvet around her. What if they get inside? What will they do? Should she barricade the door? Then she hears, from above, a skittering sound, as though something with talons is sliding around on the roof tiles. Sophie moans and dashes for the fireplace in the room, not used for decades, but still with its ash shovel, pan and poker.

Gripping the poker, Sophie stands in the middle of the room, turning a circle on her heel, looking up at the ceiling where she hears scratching sounds on the roof. They seem louder, and more plentiful. There's more than one up there. She whirls round as the sash window rattles, just in time to see what appears to be a distended, smiling mouth full of rows and rows of teeth disappear from sight.

What do they want from her?

Her heart pounding in her chest, she braves the window, and looks out. There are shapes flitting around the front of her cottage, scuttling like rats. She should have relit the lanterns. Shouldn't have fallen asleep.

There's a sudden, loud bang from downstairs. The kitchen door.

Oh God, she thinks. *They're in. They're here.*

Then she hears it, a softer sound, but rasping and dry.
Sophie.

They're calling her name.
Sophie.

More voices, inhuman, guttural.
Sophie. Sophie. Sophie. Come see.

She isn't quite certain if she hears the words, or senses them, feels them in her head. Gripping the poker tightly, she puts her hand on the doorknob. Come see what?

Sophie. Sophie. We brought her to you.

Taking a deep breath, Sophie wrenches the door open. The lights are all out downstairs, even though she is sure she left the staircase lamps on. She can sense movement at the bottom of the steps. And there's a light, pale and blue, like a television set just showing static.

Cautiously, Sophie creeps to the top of the stairs, and looks down. It's not just dark at the bottom, it's as though there's an absence of light, like it's all been sucked away. Save for the pinprick glows of seven pairs of eyes arranged around a small, mewling parcel of light.

We brought her to you.

The light resolves and sharpens, and there's a sudden cry as it fills its newly coalesced lungs with air. It's a baby, wrapped in blankets, flickering in and out like an indistinct signal.

'Take her away!' screams Sophie. 'I don't want to see her! Take her away!'

The baby gurgles.

'Please,' whispers Sophie, the strength deserting her. 'She's dead. Please. This is horrible.'

She sinks to her knees at the top of the stairs, no longer even strong enough to hold the poker, which clatters across the floor. Then the flickering light-baby seems to fade, and the darkness seems less dense, and suddenly there's a loud crash, the kitchen door slamming, and the lights fizz back on in the house.

Sophie lets out a ragged breath. They've gone. They did what they came for, and they've gone.

But gone where? If they're not in the woods, and they're not here...

She has to move quickly. She grabs a small rucksack in her room and starts to stuff things into it, clothing, trinkets, books, then stops. She does not need these things. None of them. She just needs to get out of Withered Hill.

When she opens her front door as quietly as possible, her little corner of the village is silent. Nobody is out. No people, nor the black shapes or their fizzing companions. But there are lights on in every cottage, every window. Everyone is at home. And playing host to... to those things.

Sophie will never have a chance as good as this.

She sets off at a run, kicking the turnip lantern on the path, heading straight for the trees, her feet pounding the wet grass. She bursts through the brittle, wintering branches of the outside growth, putting up her hands to shield her face, but not slowing her pace.

Over the preceding months, Sophie has become attuned to the woods, learned to sense the things in it that are not birds or mammals, to feel the presences there as though they are a weight, a pressure, she feels in her mind.

Now they are gone. Now the woods feel flat and empty and inanimate. She runs. Faster and faster, no tree branches grasping for her, no roots trying to trip her, no ivy curling around her ankles. It is dark, pitch dark, but she knows she just has to keep running in a straight line, and she will emerge at the other side of the woods, and be free of Withered Hill.

The trees thin into a narrow clearing and she picks up the pace, but her foot hits something hard, a root or rock, and she goes sprawling into the wet moss. It's fine. Just clumsiness. Not the woods trying to stop her. She picks herself up to her hands and knees and her breath catches in her throat.

He is there, seated on a hillock, elbows on his knees, scratching his chin with thin, claw-like fingers. His face is hooked and gnarled like the tree branches, but as white as the moon beneath his tatterdemalion cloak and ragged hood. It is as though he has been carved from pale, young wood and clothed in the oldest night there ever was, as though he has grown there, like a living statue, his eyes shining like fireflies framed against the blackness of the hour of the wolf. He looks simultaneously ancient and yet vital and full of life. He speaks of spring and autumn, of new growth and burnt death, like the unholy progeny of the summer and winter solstices made some kind of flesh. He is the turning of the seasons, his scent is musk and soil and sweet lavender.

She should have known. *They* might have left the woods, the rest of them, on whatever mission this Samhain night brings, but not him. Of course not him. How could she have been so stupid? How could she have thought her futile plan would work? How could she have thought that he would abandon the woods?

He, who sits in front of her, pinprick eyes piercing her soul.

He, who is known to her, whatever cellular memory she has from her time outside suddenly plucking at her nerves like violin strings and sending his name shimmering along every sinew.

He, who is the master of this place and, she now knows for certain, the master of her.

Owd Hob.

31

Inside

Days in Withered Hill: 138

'Tha's a pretty little thing,' says Owd Hob.

Sophie's eyes widen. Memories. A churchyard. A girl of twelve. Her. A ghost. She says, barely above a breath, 'You cut my hair.'

'Did I now?' says Owd Hob.

Another memory surfaces. In the back seat of a car, her parents arguing. With her. And then, ahead of them, a shape unfurling, black and white, like a pile of leaves in the eddy of a whirlwind.

'That was you.'

'It was,' says Owd Hob. 'But was it *you*?'

Sophie pushes herself back so she is kneeling before Owd Hob. 'I think so,' she says, frowning. 'Of course it was.'

'Then tha's not ready to go, not yet.'

'You mean... I'm a different person than I was when I saw you? Or I have to become a different person?' Sophie no longer feels afraid, not in the way she was when she first beheld Owd Hob. The sheer, mindless terror has subsided. Now she fears him in the way that she thinks the villagers fear him. With the respect afforded to one who is not just different from them, but *more* than them.

'You are a different person,' says Owd Hob. 'What you have to do is realise why.'

'Who are you?' says Sophie.

'Tha' should be asking that of tha'sell.'

'But who are you?'

Owd Hob stands, and his ragged black cloak seems to writhe like something alive, like the shadow of a thing that should not exist. There is illumination in that little clearing, but Sophie cannot tell the source of it. In the folds of his rags, there seem to be infinite spaces, containing multitudes. Owd Hob eats light, yet emanates it. He is like a hole cut from the air and filled with something else.

'I am both the land, and of the land,' he says. 'I am the air, and of the air. I am the water, and of the water.' His black eyes seem to glow. 'I am the fire, and of the fire.'

Owd Hob sweeps his hand and the bare, dead trees seem to shudder, and then burst into life, buds and then leaves sprouting from their dry branches, insects crawling along their moistening bark, birds singing in the exaggerated colours of their foliage.

'I am all of that, and all of that is me,' says Owd Hob. 'I was here before anything, and after everything here, I shall be. But the places like this, of folk like me, are receding, are vanishing, are being trampled upon.'

'Folk like you? There are more of you?' whispers Sophie.

'We are legion,' says Owd Hob. 'Or, we were. When this land was covered by the great forest, when all that lived did so by the whim of us, when the world was in balance. Now...' Owd Hob holds out his long, skeletal finger and a brilliant blue butterfly alights upon it. 'Now, these places are crushed, defiled, poisoned, *withered*.'

Sophie gasps as Owd Hob's other hand darts out and closes around the butterfly, squeezing tight.

'This is what man does, to all that is beautiful, all that is under Owd Hob's auspices, and the auspices of the dwindling numbers of those like him.' He holds out his white hand, the butterfly crushed and broken in his palm. Sophie feels so sad that she might cry. 'This is what man does,' he says again. 'But it is not irreversible.'

The broken butterfly shimmers, and then is whole again, and it flutters from his hand into the dazzlingly bright leaves of the trees. Sophie now feels like weeping tears of joy.

'Why am I here?' she says in a barely audible whisper.

'To help me,' says Owd Hob. 'To help me undo the damage.'

'But how?'

'Owd Hob needs a wife.'

Sophie feels her stomach turn. She feels the dark, heavy fear return. No. She does not want this. This cannot be why she is here. This cannot be her destiny.

'I don't... don't... want this. No.'

'Because you don't understand,' says Owd Hob. 'And when you do, then it will be time to go.'

There is something else in his outstretched hand now, dust or seeds or dandelion clocks, and he purses his thin white lips and blows it, right into Sophie's face.

'Forget,' he says. 'Until it is time to remember. Run, Sophie Wickham. Run.'

Sophie is running blindly through the trees, and she gasps as she emerges into the clearing near her cottage. She does not know where she has been, or why she was in the woods. The last thing she can remember is the silent procession of shapes and blurs, walking in pairs, lit by lantern-light, from the dark trees. She can see none of them, has no idea what they were or where they are.

Sophie does not want to be alone. She runs. Runs, runs, runs, until she lands, breathless, at Catherine's door.

Catherine opens it to her incessant hammering, a look of annoyance flitting across her face. 'Sophie. I told you to stay home tonight.' Then she looks down at her muddied boots, her scratched face, her torn clothing. 'What have you been doing?'

'I don't know,' says Sophie.

Catherine bites her lip. She looks past Sophie, around at the other houses and the woods beyond. 'I should send you home. I'm not sure it's safe. You'd better come in. But don't say I didn't warn you.'

Sophie tumbles through the door and into Catherine's living room, where the wood burner is blazing, and shucks off her jacket. Then she stops, eyes wide, staring.

Sitting in the chair is a young woman, younger than Sophie and Catherine. She is wearing a tea-dress and pumps, and her pale hair is set in a wave at the sides that seems out of time. But that is not why Sophie is staring.

The woman smiles at her, her face fizzing like an old television set that is on its last legs. Her entire body seems to be a projection, a pale blue image that should not be there. She says, her mouth not quite synchronised with her words, 'Hello, dear.'

Catherine is suddenly at Sophie's shoulder. 'Gladys, this is Sophie Wickham. Sophie, this is my grandmother.'

Sophie sits huddled by the fire, cupping a big glass of brandy, casting shy glances at Gladys. She looks at Catherine, who smiles.

'This is Samhain. The night when the walls between worlds are paper-thin. They bring them to us. Just for a few hours. Those we have lost. They escort them over the divide, across the black river, and we spend time together. We never know who will come, or if anyone will come at all. But mostly, they come.'

'Why?' asks Sophie.

Catherine shrugs. 'It's always been the way in Withered Hill. Perhaps as a thank you for our service. For our belief.'

'You're one of them, aren't you, dear?' says Gladys. Her appearance is of a young woman, but her voice is that of someone older and wiser. 'One of the Returns.'

'Returns…' says Sophie.

'They used to use that a lot, not so much now,' says Catherine. She smiles at Gladys. 'Ever since someone Returned a little too soon.'

'Naughty girl,' says Gladys, mock-scolding Catherine. 'That wasn't my fault, as well you know.'

Sophie looks from one to the other. 'What does that mean?'

'How long have you been here, Sophie?' asks Gladys.

'Since June.'

'Ah, well, I'm sure they'll tell you at some point. Perhaps a little early, yet.'

Sophie looks at Gladys, fascinated by the static, scratchiness of her, the image simultaneously flat and insubstantial and yet solid and rounded. She says, 'What is it like, where you are?'

'Shush,' says Catherine, laying a hand on Sophie's arm. 'We don't ask that. We merely spend time with our loved ones, just for a few hours.'

Sophie's hand flies to her mouth. 'Oh! Catherine! I'm so sorry. I'm taking up your precious time with your grandmother. I'll go.'

'Oh, I wouldn't do that, dearie,' says Gladys. 'Not when they're out there. They have their revels on Samhain, you know. While the living and the dead are cosy inside.'

'Gladys is right,' agrees Catherine. 'You'd better stay, now you're here. At least until they come to take our loved ones back.'

Sophie finishes her brandy. 'Catherine, can I go upstairs? Perhaps lie in bed for a while? Until it's safe to go? Then you can spend time with your grandmother.'

'Of course. Get into my bed. Try to get some sleep. You probably won't remember much of tonight. That's the way of Samhain. I'll wake you when it's safe.'

As Sophie stands, Gladys reaches up and takes her hand in both of hers. It feels like wool and nettles and electricity. 'Sophie, dear, before you go, just let me say this. It is early days for you in Withered Hill, relatively speaking. The answers to your questions are all in your hands already. Be patient. Let it all come to you.'

'It feels as though it will never come, Gladys.'

'It's just a case of knowing it when it does, dear. Night night.'

Sophie climbs the stairs, suddenly weary, as music bursts forth from the living room, a crackly recording. She hears Gladys applaud and say, 'Oh, Dinah Shore! My favourite!'

Sophie has slept, she doesn't know how long for, when she rouses as Catherine slides into the bed beside her. She murmurs sleepily, 'Has Gladys gone?'

'Yes, they came to take her back.'

'Was it lovely?'

'Very lovely.'

Sophie wakens a little more. 'Then I should go back to my cottage, if it's safe.'

'Hush,' says Catherine, and Sophie feels her warm body press against hers. Sophie is in her knickers and an old T-shirt she found on Catherine's floor. Catherine is naked. 'It's still Samhain.'

'You said I won't remember anything from Samhain.'

'No. You won't.'

Sophie feels Catherine's hot, sweet breath on her neck. She says quietly, 'That's a shame.'

In the morning, Sophie awakes in her own bed, refreshed and warm, with a sense that something wonderful and miraculous has occurred, but she cannot for the life of her say what it might be.

32

Outside

Days to Withered Hill: 5

On Monday night, Sophie packed what she thought she'd need for a week in Cornwall and checked the weather forecast on her phone. She had no idea whether they'd be near any beaches, or even if she'd get any free time, but she decided to throw two bikinis and a swimsuit into her case anyway. She was glad she was going with Mandy, and not Colin, she thought as she looked at the tiny Gucci bikini bottoms, then felt a flash of guilt. He was dead. It was quickly replaced by anger. He had been stalking her. It was a good job he'd killed himself, because she might have done the job for him. Or, no. Had him arrested and banged up in jail. She spent a few minutes thinking about what a terrible time he'd have had in prison, basking in a schadenfreude she'd never truly experience. He'd got away with it. But then, he was dead. She couldn't decide whether she was glad or angry that he'd escaped justice, and then it all circled back to a layer of vague guilt again.

She'd gone back to the office to finish the work she'd tried to get out of. She was glad she did. It only took her ten minutes to type in the list of words, but by the time she'd finished, the headache that was crippling her was completely gone. As was the memory of what she'd been doing when Mandy called her. Sophie frowned at the thought. Something about… She shook her head. No. It was gone, apart from a vague feeling, and the thoughts it was dredging up she wasn't particularly keen on. She let it go.

When she'd finished packing, she dragged the rug back fully over the pentagram carved into the floorboards. She was going to have to do something about that. She shivered at the thought of Colin being in here, feverishly scratching the lines into the floor, while she was out at the pub with Tom. What else had he been doing? She thought of what the policewoman had said, about him with the bag over his head, surrounded by photos of her. Sophie went to her knicker draw and looked in, her face curling in distaste. Then she scooped all the underwear up and threw them in the bin, then she went on her phone and ordered a whole new batch to come by express delivery the next day.

Then she called Tom to tell him that they'd have to put their date back. He was fine about it, which actually caused Sophie to frown. Was he not that bothered, or was he just playing it cool? She decided on the latter. He was bothered about her. She knew it. *This could be it, Sophie Wickham*, she said to herself. *He could be the one.*

She scrolled through her phone while something burbled in the background on Netflix. Tom didn't seem to have any social media presence at all. Definite man of mystery. She quite liked that, in a way – at least there'd be no suffocating attention like Jamie had lavished on her. On the other hand… someone as successful and good-looking as him must get a lot of female attention. She was almost suspicious of anyone who wasn't on social media. Did that mean there was something to hide?

Sophie pushed the thought away. *Stop meeting trouble halfway*, she scolded herself. *Just enjoy this.*

Days to Withered Hill: 4

When Sophie's delivery came the next morning, she finished packing and, to assuage her guilt at buying new clothes she couldn't afford, she decided to take some things she hadn't worn for a while to the charity shop.

Kath was working and threw her hands up in delight at seeing Sophie. 'Oh, thank God you're here. Winnie is poorly and I have to go for my blood test. You couldn't mind the shop for an hour could you?'

'I'm not here to work, just to drop these off,' said Sophie, brandishing the black bin bag full of clothes. Seeing Kath's crestfallen face, she added, 'I could do an hour, though.'

'You're a darling,' said Kath, slipping on her coat. 'Price that stuff up you bought and stick it on the rails, would you?'

Sophie spent the next twenty minutes putting tags on the clothes, dithering over whether she really wanted to part with the black top and the green skirt, and then the first customer came in, one of the regulars, an old man riffling through the rail of suits and shirts. He left without buying anything, then two women came in, one standing with her head on one side, her finger trailing along the spines of the paperback books, the other selecting two cardigans.

No one else came in for the next twenty minutes. Sophie made herself a cup of instant coffee in the little kitchen at the back and flipped through an old *Country Living* magazine from the pile on the floor by the books. She looked up as the bell clanged over the door, and a woman came in, about her age, with an expensive-cut bob and wearing a fitted two-piece suit.

Sophie's blood froze in her veins.

It was Niamh Glenister. Older, but then so was Sophie. Immaculate-looking. Beautiful. She walked straight to the desk and put a leather holdall on it.

'I've brought some clothes. Dresses, mostly. All designer labels. Hope they can be of use. I'll need the bag back. It's Louis Vuitton.' She bit her lip, then said, 'Actually, you can keep that, too. I've got another one.'

Sophie kept her gaze down, avoiding eye contact, and muttered a thank you, but she felt Niamh's penetrating stare on her.

'Oh,' she said. 'It's you. Sophie Wickham.'

Sophie forced a smile and looked up at Niamh Glenister. The woman whose life she had ruined at university.

Sophie met Niamh's unflinching gaze. It seemed as though the day had darkened a little, the sunlight that had pierced the windows before was now absent.

Niamh said again, 'Sophie Wickham.'

'I'm sorry,' Sophie blurted out, feeling eighteen again, not thirty-two.

Niamh smiled.

Not the smile Sophie remembered, the one that masked malevolence and plotting. It was a normal smile, a genuine one. Niamh said, almost conversationally, 'I cut my wrists in the bath. After it happened. Did you know that?'

Sophie nodded, her eyes dropping to Niamh's hands, still holding open the bag full of designer dresses. Niamh followed her gaze, and turned her hands palm upwards, so Sophie could see her wrists.

'All healed,' she said, smiling again. 'It was a long time ago. I won't pretend that it wasn't bad for a while. I went off the rails. Did some bad things myself. Things I'm not proud of.' Sophie met her eyes again. 'I was studying law, do you remember? I wanted to be a lawyer. If you hadn't confessed, I'd never have been allowed to do it. Not if I got a criminal record.'

'Did you stay at uni?' said Sophie quietly.

Niamh shook her head tightly. 'No. Not after what happened. I spent some time in a mental health unit. Private, of course. My parents were very well-off. Then, for a couple of years... well. I got into drugs. Fell in with a bad crowd, as they say. Then something happened in my early twenties.'

'What happened?'

'I became a different person,' said Niamh with a smile. There was something about her eyes. Something shining there. Almost like fervour.

A born-again Christian, Sophie thought.

As if reading her mind, Niamh laughed. 'Nothing like that. Though... perhaps. In a way. But nothing to do with the Church. I just became a *better* person. The best Niamh there could be. And I went to another university, and I did study law, and, well...' She smiled. 'Now I'm very well-off all in my own right.'

Niamh looked at her curiously for a moment, so intensely that Sophie began to feel uncomfortable. Then the other woman took hold of Sophie's wrists, gripping them tight, so tight it almost hurt.

Niamh whispered, 'You're going to become a better person too, Sophie Wickham. I can see it in you. You've been marked out.'

Sophie politely tried to extricate herself from Niamh's grip, but she tightened her hold. Her smile looked suddenly false and forced.

'Perhaps we can spend time together, Sophie, when you *are* a better person.'

'You're hurting me,' said Sophie. Light seemed to be leaking away, darkness crowding the periphery of her vision.

Niamh said, 'Oh, this is nothing compared to what's to come.'

Then the bell tinkled above the shop door, and the darkness outside seemed to lift, and sunlight flooded through the windows again. It was Kath, returning from her appointment.

Niamh let go of Sophie's wrists, smiled at her, and walked out without another word.

Kath began to sort through the bags, cooing at the labels on the dresses. 'Ooh, these are lovely! Did you ask her if she wanted to Gift Aid?'

Sophie said nothing, just rubbed her wrists and watched the figure of Niamh Glenister disappearing across the street.

33

Outside

Days to Withered Hill: 4

'Who was that?' said Kath, holding up one of the dresses admiringly. 'She seemed to know you.'

'Someone I was at university with,' said Sophie absently. What had Niamh meant? She would see Sophie again when Sophie was a better person?

'Ooh, wish I'd gone to university,' remarked Kath, clucking her teeth. 'All parties and boys and wacky baccy I bet.'

'Something like that,' replied Sophie.

—

When she first got to university, Sophie didn't know anyone – nobody knew anyone, of course. And, like a lot of people, she decided that this was the perfect opportunity to reinvent herself. Sophie Wickham until the age of eighteen had been unremarkable. Boring. A bland nobody from a bland Cotswolds village with nothing to say or do to make her interesting. A dull virgin with little or no redeeming features. That was going to change. And from the moment she swept into her halls of residence, leaned on the doorframe of the boy next door who was unpacking his things and declared, 'Right, are we going to the fucking pub or what?' the new Sophie Wickham was born.

The next fortnight passed in an alcoholic haze. She found she was funny when she was drunk. She was attractive. She shed her inhibitions and danced and laughed and felt utterly free and minted afresh. This was who she was. This was who had been hiding inside her all the time.

And then Niamh Glenister came along and spoiled it all.

Niamh had missed freshers week because she had been volunteering over the summer with a water charity in an African village, building a well. She was studying law, confident and breezy and everyone loved her immediately. And she moved into the empty flat in Sophie's block, right next door.

The carefully constructed facade of the new Sophie Wickham seemed to crumble under Niamh's piercing blue-eyed gaze. It was as though the new girl saw straight through her, saw who she really was and what she was pretending to be.

'Hello,' said Niamh, by way of introduction as she unpacked her designer luggage. 'Who are you, next-door neighbour?'

When Sophie introduced herself – all the bravado she'd built up over her first couple of weeks at university seemingly deserting her – Niamh had laughed delightedly.

'Sophie Wickham,' she said, copying Sophie's accent.

Later, when she'd settled in, she'd knocked on Sophie's door and said, 'We're all going to the pub, Bumpkin, are you coming?'

Normally, it was Sophie who corralled everyone to go out. Niamh had been in the block for five minutes and everyone was fawning over her, deferring to the self-confident superiority that she seemed to naturally exude, born of money and breeding.

Sophie bristled at the nickname, and immediately began to modulate her West Country accent, smoothing it out to be more like the way the rest of the girls, all from around London, spoke.

Everyone else in the block adored Niamh, and she quickly became not just part of their tight-knit circle, but established herself at the top of the pecking order. Sophie and Niamh circled around each other like sharks, wary, competitive, giving each other respectful space while maintaining a facade of friendship. Until one night.

There was a party, a packed student house heaving with bodies, dance music playing in the darkened rooms. Sophie had gone with her flatmates but lost them almost immediately, so danced alone, feeling hungry eyes upon her. She liked the feeling. She had yet to sleep with anyone at university, had yet to give up her virginity, but there was no rush. She was just waiting for the right time.

Someone pressed a bottle of beer into her hand and she glugged it thirstily, feeling the press of bodies against her. A hand gripped her hip. She turned round and there was a big guy there, one of a bunch of medical students she'd noticed earlier. He leaned forward and said something she couldn't hear. Instead of replying, Sophie danced, holding the bottle in the air. He stayed by her side. Sophie was reeling him in, gently, skilfully. She had already decided he would be the one, and tonight.

He was called Neil and he was quite lovely. They talked animatedly about movies and books and travel and music, each topic punctuated by a bout of kissing in the kitchen. He was from Bristol and played rugby, though he confessed he didn't really like it much. He was studying medicine because his father was a doctor, but what he really wanted to do was sit on a beach in Goa and write poetry while the sun set over the iridescent ocean. Sophie thought she might be in love.

At some point, the party broke up and a bunch of them went back to their halls, continuing the revelries in their kitchen. As dawn was breaking, there was only Sophie, Neil and a couple of others sitting around the table, drinking what they had left. And Niamh.

Sophie kept stroking Neil's hand, impatient to get him into her bed. But like a moth to a flame, his attention kept being drawn to Niamh, luminescent even as Sophie felt she must look like something the cat had dragged in by then.

Niamh was mixing cocktails with the dregs of the booze, and they were giving them stupid names. 'I call this one… the Bumpkin!' Niamh had said delightedly, pushing a foul-tasting pink concoction in front of a scowling Sophie.

And that was the last thing she remembered.

Sophie awoke with a splitting headache and a mouth like the Sahara, groggy and unsure of how she had got into bed. There was a rhythmic squeaking of bed springs from the room next door, and the mingled sounds of sex. Sophie staggered out of her room to get water, and Niamh's door was wide open. She stopped and stared at Niamh straddling Neil, and the girl looked over her shoulder, locking eyes with Sophie, triumph on her lips.

Sophie ran to the kitchen and threw up in the sink, and as she was running the tap into a pint glass, she saw, among the detritus of the bottles and mixers, an empty blister pack of pills.

Later, as Sophie sat in the kitchen swaddled in her dressing gown and coolly watched Neil sneak out of Niamh's room and wretchedly scuttle off into the morning, she tossed the empty packet across the table as Niamh came in to make a cafetière of coffee.

'These yours? Alprazolam?'

Niamh shrugged. 'They help with my anxiety.'

'Yes, I googled them,' said Sophie, not believing for one minute that Niamh Glenister was ever anxious about anything. 'They're a sedative, as well.'

'Yes, good for helping you sleep.' Niamh sat down with the coffee, across the table from Sophie.

'You put them in my drink,' said Sophie levelly.

'Why would I throw away good pills on you?' said Niamh with a snort.

Sophie looked away. 'So you could fuck Neil.'

Niamh laughed and picked up her mug and the cafetière. 'I'll take this in my room. There's too much trash in here.' At the door, she stopped. 'I've finished with him, anyway. He's all yours.'

Anger flashed through Sophie. 'Do you always just take what you want, Niamh?'

Niamh seemed to consider this. 'Mostly people just *give* me what I want. But yes, if I need to, I just take it.'

Sophie fumed alone in the kitchen, revenge uncurling in her head like an awakening serpent.

–

'Has anyone seen my bracelet?' said Fiona, scouring the unopened mail and random collection of phone chargers and cups scattered on the kitchen work surface. 'I took it off and put it on here when I washed up this morning.' She cast a critical eye over the kitchen. 'Which, apparently, needs doing again already.'

No one had seen Fiona's bracelet. Just like no one, over the next two weeks, had seen Georgie's earrings or Katy's watch or India's locket containing a photo of her granny or Sophie's gold ring. Everyone seemed to be losing things. Books, bank cards, photo frames, keys. Small things that people didn't notice at first until they needed them or remembered they'd put them down somewhere. Everyone in the flat was missing stuff. Everyone except Niamh.

Still, Sophie was a coward. She knew she'd crumble under Niamh's denials and subsequent inevitable cross-examination. The girl had a stellar career as a lawyer ahead of her. It was in her blood. But an opportunity presented itself when Niamh left the university early for the Christmas break, to go skiing with her family.

The day after she'd gone, Sophie made a big deal of having lost her ring, saying it was the only thing she had to remind her of her dead mother. She managed to get everyone else fired up and Fiona, ever the stickler for the rules and doing what was right, suggested a room-by-room search. There was a thief in their midst, and it had to end now.

Sophie willingly opened up her room for the search, and when all the girls had done the same, they stood outside Niamh's door. She hadn't locked it. Niamh had the most expensive clothes and jewellery and tech, but she didn't care if she lost it. She'd been unconcerned by the items going missing; she could always get more.

Sophie hung back as Fiona led the search and quickly came across the box that Sophie had hidden right under

Niamh's bed, behind her suitcases, against the wall. The box containing all their missing things.

'Bitch,' said Georgie.

'I'm going to report her,' said Fiona. Years later, Sophie would hear that she'd joined the Metropolitan Police as a detective.

A couple of days before they all went home for Christmas, Fiona called a house meeting. Ashen-faced, she said that she had been told in the strictest confidence that the university had contacted Niamh and her parents in Val-d'Isère. Niamh had protested her innocence vehemently, begging them not to involve the police, or her law career was over before it had begun. The university had said that it was up to the victims of the crime if they wanted to press charges. Regardless, Niamh would not be welcome back at university.

'So, what do we do?' said Sophie. She felt a little sick at the thought of the police being involved. Wouldn't her prints be all over the stolen items? She hadn't intended for it to go this far.

'I think things have overtaken that slightly,' said Fiona in a hushed tone. 'Yesterday Niamh slit her wrists. In the hot tub at the chalet. On the mountain.'

'She's dead?' said Sophie hollowly, darkness crowding the periphery of her vision.

'No. They got to her in time. But, still. Apparently she suffered from crippling anxiety. You'd never have guessed to look at her, would you? She was on medication.'

They decided unanimously that, although Niamh was indeed a bitch and a thief, she had probably been punished enough, and they would not press charges.

Sophie excused herself and walked hurriedly out of the block, quickening her pace until she was running to the small copse of trees that bordered the campus, where she leaned on a tree, hyperventilating.

This was the worst thing she had ever done in her life.

Is it though?

The words seemed to be in her head, as well as floating around her.

She looked up sharply, and caught a flash of something in the trees. Something white and earthy and bone-like, ragged and torn, there one second, gone the next. Something that tugged with hair-thin filaments at a memory from when she was little. And another memory, from when she was even younger, that she had locked away in a box in her head, just like she'd locked away all her friends' possessions in a box under Niamh's bed.

Feeling sick to her stomach, Sophie went to the vice-chancellor's office to confess everything.

34

Outside

Days to Withered Hill: 4

Mandy asked Sophie to meet her at the office before they set off for Cornwall, and Sophie wheeled her case through the doors and took the lift to their floor. Mandy was waiting for her, with her own case, tapping away at her keyboard in her office. She waved Sophie in and closed down her computer.

'I've ordered us a car to take us to Paddington,' she said, looking at Sophie's case, twice the size of hers. 'We're not going on holiday, you know.'

'We're going to Cornwall in June,' replied Sophie. 'It's the nearest I'll get to one this year.'

Mandy shrugged and held out her hand. 'Have you brought a laptop or tablet or anything? I'll need them, along with your phone.'

Sophie looked stricken. 'Phone?'

'It has to stay here. Nature of the job, I'm afraid.'

Sophie pulled her phone from the pocket of her light jacket and looked at it longingly before handing it over. How was she supposed to contact Tom? How was she supposed to contact anyone? She said, 'What if there's an emergency?'

Mandy gave her a thin-lipped smile. 'There won't be. There's a landline, anyway.'

'Are we travelling back in time for this job?' muttered Sophie.

'In a way, yes, we are,' said Mandy.

—

The train was bound for Newquay, which brightened Sophie up a little. That was busy, wasn't it? Big holiday town? Beaches? She was glad she'd packed the swimwear. Maybe this wasn't going to be so bad after all. The journey took five hours, which she wasn't relishing; she'd never been in Mandy's company alone for so long. She wasn't sure what they'd talk about. Certainly not the encounter at the shop with Niamh Glenister. Sophie was still unsettled by that, and had already decided that whatever Niamh said, however much time she'd spent getting treatment after trying to take her own life, it hadn't been enough.

In the event, she needn't have worried. Mandy had booked them a table seat in first class, and as soon as the train pulled out of Paddington, she took a book from her bag and started to read.

Sophie knotted her fingers in her lap and stared out of the window, feeling as though she'd forgotten something. She fiddled with a piece of thread on her trousers, patted her pockets, drummed her fingers on her knee. She caught Mandy staring at her over the top of her paperback.

'Are you really so wedded to your phone that you can't go a few days without it?'

Sophie looked down guiltily at her hands, her thumbs moving of their own volition, tapping against an invisible screen. They'd only been on the train half an hour. How was she going to last a week?

'Didn't you bring a book?' said Mandy.

Sophie glanced at the paperback she was reading. The cover showed a woman, shot from below, framed against a blue sky, two dark birds high above her. 'I brought three. Is that a book of tips on how to win the lottery?'

Mandy frowned, then gave a sharp exhalation of air, which Sophie realised was a laugh. She said, 'You've never heard of Shirley Jackson?'

Sophie shrugged.

Mandy put her bookmark in the page and closed the book. 'You can read it while we're there. This is a book of short stories. *The Lottery* is about a town where they draw lots every year and the winner… well, the loser is stoned to death.'

'That's horrible,' said Sophie. 'Why?'

'To ensure the future prosperity of the community.'

'And what happens? Does somebody rescue the person in the story?'

'No,' said Mandy. 'They die.' She looked at Sophie for a long time. 'That horrifies you? That one person should be sacrificed for the greater good of everyone else?'

'Yes!' said Sophie. 'Doesn't it horrify you?'

'It does,' replied Mandy, seemingly satisfied that Sophie had given the right answer. She felt like she had passed some kind of test. Mandy leaned out into the aisle. 'The steward is coming. Shall we get some coffee and brunch?'

-

As the train entered the Cornish peninsula, Sophie found herself reaching for her non-existent phone less and less, losing herself in the countryside whipping past the window. She pulled a face at a wind farm on the hills, the ranked, silent turbines rotating slowly in the summer breeze. A field they passed was not given over to crops or grazing, but rows and rows of solar panels. Sophie noticed Mandy scrutinising her.

'You don't approve?' asked Mandy.

Sophie shrugged. 'I don't care one way or the other. But none of it's exactly pretty, is it?'

'That doesn't make you a bad person.'

Sophie frowned. That seemed like an odd thing to say, but she didn't pursue it. Bored of the view, she said, 'Where are you from?'

'London,' replied Mandy. She put her book down. 'I say that because you're not from London. Bromley, specifically.'

'I've lived in London long enough. I know where Bromley is.'

'You're only passing through London, no matter how long you've lived there, unless you're born in London.'

Sophie looked at Mandy. She could do so much more with herself. She wasn't bad looking, but her hair was scraped back and she wore no make-up. A bit of colour could liven her up no end, give a bit of volume to those bloodless lips. Maybe Sophie could give her some tips this week, as they were thrown together on what was shaping up to be the worst girls' holiday ever.

'How long have you been with the company?'

'Few years.'

'Why did you join?'

Mandy looked at her. 'Because I wanted to help people. And there are... family connections.'

Sophie tried to stifle a laugh. 'You joined a data-processing firm because you wanted to help people?'

Mandy picked up her book again. 'You'll understand, one day.'

Sophie thought about all the meaningless numbers and words and data she'd inputted in her time there. Who was that helping? Well, she supposed people wouldn't pay for it to be done if it wasn't useful. But she'd never thought of it as *helping* anyone before, and still didn't. It was a job. She got a wage for it. And that was about as far as it went.

The guard announced that the next station was Newquay and Mandy started packing her things into her bag. As they waited for the train to pull into the platform, Sophie caught a glimpse of people thronging the streets, shorts and T-shirts and flip-flops, pennants strung between the shops, fluttering in the June breeze.

She turned to Mandy in the aisle. 'Are we staying in Newquay?' she said hopefully.

'A little outside,' said Mandy.

The train hissed to a halt and disgorged the passengers, and Mandy led her through the ticket barrier to the road. A silver Volvo was waiting for them, a man in a black suit and tie nodding as he saw Mandy.

'Miss Scott,' said the driver, a wiry, thin-haired, bespectacled man who must have been knocking on seventy. It sounded funny, someone calling Mandy 'miss'. Sophie had never wondered if she was married or even with someone; she'd never wondered much about Mandy at all, to be honest. 'Good to have you with us again,' the driver remarked as he loaded their cases into the boot. 'You've picked a fine week for it.'

'We won't be doing much sightseeing, Andrew,' said Mandy. 'How are the twins?'

'In good health,' replied Andrew. Sophie listened to the conversation silently. Twins? Andrew opened the door for Mandy. 'I was terribly sorry to hear about Colin.'

Mandy nodded and said, 'Andrew, this is Sophie. She was originally meant to be coming with Colin. I stepped in at the last moment as it's such an important job.'

Andrew looked over the rim of his glasses at Sophie with interest. 'Ah, yes. Miss Wickham. I very much hope you'll enjoy your stay at Macha.'

Sophie glanced at Mandy, who said, 'Macha House. Where we will be working for the rest of the week.' She slid into the front passenger seat. 'Come on, Andrew. Time's a-wasting. I'd like us to get at least an hour or two of work in before dinner.'

As the car navigated the narrow streets to the outskirts of Newquay, Sophie, from the back seat, watched the glittering sea and guessed she would not be making much use of those bikinis at all.

Andrew directed the car along the coast, past picturesque villages where grey stone church spires stained with lichen rose up beside inviting pubs, and Sophie wondered what Tom was doing. She had phoned him on Monday to cancel their Wednesday date, telling him she had to go away for work.

'I know it sounds ridiculous, but I can't tell you where,' she'd said.

'Right,' he'd said, a little tightly in her opinion.

'This isn't me making an excuse,' she'd said hurriedly. 'I'm gutted that we can't meet up again on Wednesday. But I'll be back at the weekend. Next Tuesday good for you?'

'I'll check my diary,' he'd said. 'Call me tomorrow?'

The car pulled in tight against a gorse hedge to let a tractor past on the narrow lane. She wouldn't be able to call Tom tonight, or any other night that she was in Cornwall. But Mandy had mentioned a landline; Sophie had already committed Tom's mobile number to memory.

Actually, she thought, it might do him good to wonder about her a little bit. Tom was all mysterious about his work, and Sophie had already felt a little mundane next to him. She'd hinted on the phone about having to sign the Official Secrets Act without actually saying the words, and he had sounded rather impressed. As, to be honest, was she. Now the disappointment at having to go away had settled, it was a bit thrilling, being driven off to some remote location in Cornwall to carry out top-secret work. Maybe she was a spy, even if she didn't know it. Perhaps what she would be doing this week was some kind of antiterrorism work, or something that would save the lives of hundreds of people somewhere.

If Tom was feeling a little put out by her call, he'd evidently mellowed, because half an hour before she'd left to meet Mandy that morning a courier van had pulled up and she had to sign for a package. She'd already had her new knickers delivered so had no idea what it could be. Ripping it open in her kitchen revealed a soft toy, some kind of… elf, she'd guessed. Or pixie? Maybe from a movie or kids' cartoon? She didn't know. There was a printed note that read *To keep you company until you see me again. Tom xxx*. She'd smiled and pushed it into her suitcase.

Sophie had no idea how long they'd been driving, but they hadn't passed through a village for a while. Mandy and Andrew had been chatting in low voices throughout the journey while Sophie looked out absently at the scenery. It was very beautiful here, and they'd kept the sea on their right side for most of the journey as they headed west. Then Andrew slowed and turned off the road, heading towards the coast. Sophie craned her neck to see through the windscreen between Andrew and Mandy, and noted they were on little more than a dirt track, threading between wide, open moorland. Ahead of them lay the blue sky and a thread of grey, sparkling sea beyond what she got the sense must be a high cliff dropping to the ocean below.

And standing in splendid isolation was a house, clinging to the edge of the cliff like a dark, foreboding seabird, large and rambling and ramshackle. Its wide roof was bowed, and its chimneys seemed haphazard and randomly placed, its windows offering no sense of balance or planning, but dotted into the dark grey slate exterior as though at the whim of the builders, not the design of an architect. In fact, as they drew closer, Sophie felt that the whole house was composed of bits and pieces of other buildings that had been dropped in the same space and scooped and pushed together as a baker might knead dough.

She leaned forward for a better look. 'This is where we're staying?'

Mandy glanced over her shoulder. 'Yes, this is Macha House. Your home for the week.'

35

Andrew parked the Volvo on the gravel circle in front of Macha House's big double doors, and opened the doors for Sophie and Mandy to get out of the car. He led them to the entrance as Sophie looked up, squinting against the late-afternoon sun, at the windows, all blankly reflecting the flawless sky. The house seemed to... lurk, hunched over as though it itself was peering back down at her, regarding her with interest. Sophie wondered vaguely just how stable the construction was, and its age.

'I'll bring your things up to your rooms,' said Andrew, letting them into the cool, dark entrance lobby. The heels of Sophie's court shoes tapped and echoed on the tiled floor. She glanced around at the dark, wood-panelled walls, adorned with paintings of stern-faced people and landscapes in muddy oils, and at the sweeping staircase that rose up from the lobby, separating left and right into tributaries that flowed up and behind onto the next floor. 'Do you wish to rest? Or you mentioned working...?'

'I think we'll take a few minutes to freshen up and then do an hour or so in the study before dinner,' said Mandy. 'I'll show Sophie up to her room if you can fetch the bags.'

'Very good, Miss Scott,' said Andrew with a nod of his head. 'Does lamb suit for dinner? The twins always have lamb on a Wednesday.'

Mandy smiled. 'I remember. Yes, I'm sure lamb will be fine...'

She glanced at Sophie, who shrugged. 'Sure, sounds good.'

'Excellent,' said Andrew. 'Dinner at six, then? In the dining room. I'll bring your things.'

Mandy led Sophie up the staircase, keeping to the left. Sophie said, 'Does Andrew do everything around here?'

'Pretty much.'

'There's no one else here but us and him?'

'And the twins.' Mandy took Sophie into a gloomy corridor lit by flickering electric lamps that looked like they were installed sometime around the First World War.

'And who are the twins, exactly?'

Mandy stopped outside a dark wooden door and turned the brass handle. It opened into a large room dominated by a four-poster bed and a floor-to-ceiling window affording views of the moors and the road they turned off in the distance. There was a mahogany desk and a washbasin, and a tall, stout wardrobe. Mandy opened a door showing a bathroom with a toilet and roll-top tub with a rubber shower attachment fixed to the taps. 'Not many mod cons here, I'm afraid.'

Sophie walked into the room and perched on the edge of the firm mattress, running her hand over the tied-back drapes around the bed. 'And the twins?'

'I'm sure you'll meet them this week.' There was a cough and a polite knock from the corridor. 'Ah, Andrew, come in.'

Andrew brought in Sophie's case and deposited it near the desk.

Mandy said to her, 'I'm just down the hall.' She looked at her wristwatch. 'It's almost four now; I'll see you in the lobby at half-past, all right?'

Sophie heaved her case onto her bed and unzipped it, taking out the plush elf that Tom had so thoughtfully sent. She placed it between the plump pillows on the double bed, and smiled. It was only a few days. She'd see him next week. Sophie hung her clothes in the musty-smelling wardrobe, worried for a moment about moths, then put her underwear into one of the drawers. She held up her bikini ruefully. It didn't look like she was going to need it. Standing at the window, she had to admit it was a beautiful, dramatic view. The sea was behind the house; maybe she could go for a walk after dinner. And, you never knew, it might be warm enough to swim.

Sophie sat on the toilet, unused to not having her phone to scroll through, then had a quick wash at the wide sink. She'd try to get to grips with the primitive shower set-up between whatever work Mandy was insisting they do and dinner. Then she let herself into the gloomy corridor and felt her way back along the way they had come to the grand staircase.

The study was at the rear of the house on the ground floor, its large windows showing sloping gardens that ran downwards to a tall brick wall bordering the house's territory. Beyond was the sea, and a distant tanker moving at a stately pace along the horizon. The room was lined with shelves holding what Sophie thought of as old books – big hardbacks with dusty green or red bindings. There was a fireplace, stacked with dry wood and kindling, with two battered easy chairs either side of it. There was a bureau at the windows, and a large oak table in the centre of the room, where Sophie and Mandy sat facing each other, laptops open in front of them.

'We have Wi-Fi?' asked Sophie, sitting down in the hard chair in front of the laptop. A wire in-tray held a sheaf of papers that at first glance appeared to be filled with tightly handwritten Latin words.

'No,' said Mandy. 'We don't.'

Sophie frowned. 'Then how do we send the work?'

'You don't need to worry about that. If you could do at least the first ten pages, please.'

Sophie set to inputting the data, casting occasional glances up at Mandy, who was tapping at her own keyboard. She didn't seem to be processing any data, so what was she doing, if there wasn't any internet connection?

Sophie redoubled her concentration; the work was making her eyes sting and seemed to be robbing her of energy. She wondered what the words meant, and why they were so top secret. Some kind of code, surely; something to do with intelligence, of course. Who had written them? Some agent somewhere deep within enemy territory? Sophie frowned to herself. Who exactly were the enemy, these days, anyway? She made a mental note to take more notice of the news when she got home.

It had just gone a quarter past five when Sophie announced that she had done ten pages. She had a bit of a headache.

Mandy said, 'Do you want to freshen up before dinner? You don't have to dress up for it. It'll just be us.'

'Not the twins?'

'They mainly keep to the top floor. Andrew will be serving them their dinner now; they tend to eat early and retire by nine.'

Sophie had questions, lots of them. Were these twins part of the company? Was it just that their house was used for this secret work, because it was so remote and seemingly unconnected to the modern world? Who was Andrew, exactly?

'Andrew is one of the twins' nephews,' said Mandy suddenly. 'He looks after them. And us.'

'Surely he's both of the twins' nephew?' said Sophie. 'Unless… he's the child of a sibling of a spouse of one of the twins.' The mental gymnastics made her head hurt more. 'Or one of the twins' son?'

'Only one twin was married, but not…' said Mandy. She closed the lid of her laptop. 'It's complicated. You don't need to worry about it. See you back in the lobby at five to six.'

–

When Sophie got out of the bathroom, after holding the grey shower pipe that dribbled hot water over her while standing in the bath, the blue sky had paled somewhat, the sun marching onto the west out of sight. She changed into a fresh blouse and some black trousers, and fiddled with the archaic plug socket to get her hairdryer and straighteners to work. The old house's ancient electricity system seemed to protest at this intrusion of the modern world, and the hairdryer kept cutting out as she was using it. There was no TV in the room, nor even a radio. Mandy had been right; it was like going back in time.

Mandy was waiting for her at the bottom of the stairs and led her into the dining room, a vast space that held a long table, two places set at the far end near the window looking out onto the gardens. Like the other rooms, it was panelled in dark wood that soaked up the meagre light from the pale lamps. Everywhere was so gloomy in Macha House.

Andrew appeared and poured each of them a glass of red wine without asking if that's what they wanted. The wine was warm and tart, but Sophie didn't care. Andrew refilled her glass when he returned with a small plate of some kind of shellfish as a starter, and Mandy told him to leave the bottle and perhaps open another one. Sophie glanced at her; was that some kind of jibe?

'So what will we be doing here all week? More of the same?' she asked.

'Pretty much,' said Mandy, dabbing at the corners of her mouth with her cotton napkin as she finished her starter.

Sophie couldn't see what was so special about the work that it couldn't be done in London, but she said nothing, just thinking about the two months' pay she was getting as a bonus.

'Do you come here often?' said Sophie, trying to make a joke of it.

'On occasion,' replied Mandy.

Andrew entered with the main courses and took away the empty wine bottle that Sophie had mostly drunk herself. He replaced it with a fresh one and Sophie fell on the meal, surprised at how hungry she suddenly was. The lamb was perfectly cooked, on a bed of crushed potatoes and summer vegetables, drizzled with a thick, minty sauce.

'Andrew is an amazing cook,' she said through a mouthful of lamb. She washed it down with a big swig of wine. 'He could get a job in a London restaurant with this food.'

'He's happy enough here, looking after the twins. He gets plenty of reward.'

After the lamb, there was a light summer fruit pudding, then coffee. The light from outside had dimmed considerably, like the dark surfaces of the room swallowed it up. Sophie said, 'I'd like to take a walk outside, if that's all right?'

'I'll come with you,' said Mandy, dropping her napkin on the remnants of her pudding. 'Show you the bounds.'

The warm wind whipped Sophie's hair across her face as they stood at the edge of the cliff. Mandy had taken her through the grounds to the boundary wall, letting them out through a faded, rotten gate. The sea crashed and roiled on the rocks more than a hundred feet below, surf smashing onto a tiny strip of pebble beach.

Mandy said, 'I love it here. The drama of it. Nature unfettered.'

'I guess I won't be doing any swimming,' muttered Sophie.

Mandy moved off to the left, leading her around the walled garden to the raw moorland that surrounded the house. She made for what Sophie realised was a standing stone, leaning and half-buried in the earth. As she got closer, she realised that it had faint, weathered carvings on it, circles and whorls. Mandy laid her hand on it. 'No one knows how old these are. But they mark the boundaries of Macha House.' She pointed to another stone, a hundred feet distant, and walked to it, laying her hand on it as well. There were more, in a big circle around the house. Mandy and Sophie walked to each one, and Mandy said, 'Don't go beyond them. Don't cross over. Not once. Do you understand me? It's very important.'

'Uh, sure,' said Sophie, uncertainly.

They completed a full circuit of the half-buried stones, bringing them back to the cliff edge and the gate into the walled garden. The light was diffuse and failing as the sun sank.

Sophie said, 'What do we do now?'

'I don't know about you, but I'm tired,' said Mandy as they went back into the house. 'I'm going to bed. You should too.'

At the bottom of the staircase, Sophie said, 'Do you think it would be all right to borrow a book from the study?'

'I thought you'd brought three?'

'I don't fancy any of them. Thought there might be something here. Something Cornish maybe.'

Mandy smiled. 'Good idea. I'll see you down in the dining room for breakfast? Eight sharp?'

Sophie nodded and when Mandy had ascended the staircase, she let herself into the study, grabbing a book at random from the nearest shelf. She wasn't there for a book. When she had been working, she'd seen a telephone on the bureau, an old black Bakelite rig with a twisted cord flex. She closed the door to the study quietly and picked up the receiver, mentally reciting Tom's phone number.

There was nothing but dead silence.

'I'm afraid the telephone connection is very intermittent.'

Sophie jumped, letting out a little scream. It was Andrew, standing by the open door. How had she not heard him come in. She composed herself and said, 'I just wanted to… check on my Gran. She's been unwell.'

'Maybe tomorrow the lines will be up,' suggested Andrew.

'I'll try then,' said Sophie, forcing a smile.

She rushed too quickly to the door and Andrew said, 'Miss Wickham?'

'Yes?' she said, turning to him. Did he know, somehow, that she wasn't calling her gran at all?

'You forgot your book.'

Sophie nodded and took it from him, then let herself out of the study, heading quickly up the stairs and glancing back down as the staircase turned to see Andrew staring impassively up at her from the tiled lobby.

36

Outside

Days to Withered Hill: 3

Sophie dreamed of Tom. She was in the crashing surf off the coast, deep in the grey, rolling waves, the sunlight lancing through from above. But she wasn't drowning, or even struggling. She was like a mermaid, drifting below the surface. And Tom was there, too. But distant, at the edge of the black depths of the sea. He kept trying to swim towards her, but the more he did, the further away he seemed to get. There were fish darting between them, brightly coloured and swift, but then they weren't fish at all, they were little shoals of Latin words. Sophie opened her mouth to call Tom's name, but all that came out was a formless scream in a constellation of bubbles. Then he was tugged backwards, as if by an invisible force, and he disappeared into the black depths.

Sophie was sweating when she woke, the plush elf clutched in her hands. She was disorientated at first, under the blankets on the wide, hard, four-poster bed, then she remembered where she was. Macha House. The dream clung to her like a patina, and she stripped out of her shorts and vest and stood under the makeshift shower until she had washed it away as best she could, then dressed and went down for breakfast.

Sophie worked until lunchtime, inputting the final pages of the sheets of handwritten Latin. The work had made her feel a little odd, as though her senses were dulled. She felt disconnected from everything outside Macha House, as though the world was receding. She realised that she hadn't even absently reached for her phone all morning.

'I think I need some air,' she said to Mandy.

Mandy looked at her watch. 'We can break for lunch now.' She considered the pile of pages. 'Actually, if you've finished, we can call it a day.'

'I'm not sure I want lunch, after that huge breakfast,' said Sophie. 'I might just go for a walk.'

'Don't go out of the grounds,' warned Mandy. 'Should I come with you?'

'I'll be fine.'

Sophie went up to her room to change and on the spur of the moment decided she might go for a run rather than a walk. She'd brought her running kit with her, and a full week of Andrew's cooking filled her arteries with despair. She changed and was sitting in the window, pulling on her shoes, when a movement flitted at the periphery of her vision.

Sophie stood and looked out of the window, to the left where she'd thought she'd seen something. There was only a seemingly endless stretch of moorland. She never even saw a car on the road they had come in on. Perhaps it was a gull, or a fox.

She went down the stairs and saw Andrew, standing awkwardly by the main doors, glancing into space and suddenly rousing himself, as though he had just woken up, or was an actor who had heard the command to move to action. He looked her up and down.

'I'm going for a run,' she said, feeling the need to unnecessarily explain herself.

He frowned as though he could not understand why anyone would go outside just to run.

'Your amazing meals,' said Sophie, laughing and patting her stomach.

He frowned again, then turned without a word and went towards the stairs.

Sophie let herself out into the garden and went through the ramshackle gate to the clifftop. The sea was calmer today, the tide was out, and peering over the edge, she could make out the wider expanse of the shingle beach. She wondered if there were caves down there, if smugglers had used them, or if wreckers had lured ships onto the jagged teeth of the black rocks poking up through the low-tide waters.

Sophie did some stretches and limbered up, and just before she set off towards the first of the sunken stones, she glanced back up at the house, at the room she was occupying. In the window above hers, on the upper story of the house, she saw a figure, pale and almost insubstantial, watching her. Then she blinked, and it was gone. Andrew? Mandy? Maybe even one of the mysterious twins?

Sophie did two circuits of the ring of stones, and stopped by the gate on her second one, hands on her knees, breathing hard. That should mitigate at least half of the breakfast she'd had, she thought. She was going to go back in and shower when she changed her mind, and walked over to the first stone again. She put her hand on it, as Mandy had done, but it felt just like a normal piece of stone. Her fingers traced the weather-faded rings and whorls, and she wondered what they meant.

Sophie looked beyond the stones, along the edge of the cliff, and frowned. It seemed... different on the other side of the invisible boundary marked by the stones. She looked back at the house, then out again. The sky was blue and vibrant when she looked up, but pale and diffuse when she looked outwards. The grass beneath her feet was green and lush, but just on the other side of the stone, it seemed... drier, greyer. It was as if there was a filter on Macha House, making it seem more... vivid.

She jumped as she caught a movement again. In the long grass outside the boundary, a shape shivered, then put up a black nose. It was a fox. It stood up straight, and looked at her, tilting its head on one side.

Sophie slowly sank to her haunches. 'Hello, boy,' she said. She didn't know if foxes would bite you, or if they carried disease. She didn't much care; she suddenly felt as though she needed some connection with the outside, whatever it was. She held out her hand and rubbed her thumb with her fingers. 'Hello, then. Come here, why don't you?'

The fox took a tentative step forward, then sat down, like a dog.

'Come on, then,' said Sophie encouragingly. She wished she had a bit of food to entice it.

The fox took a step forward, and another, until she could almost reach over the boundary line and touch it. Then it sniffed the air and looked to the right and left. It inched forward a little more, and paused, putting its head down as though to butt something invisible. Then it looked up at Sophie with an almost quizzical expression, turned around with its bushy tail held high, and ran away from her, into the long grass.

Sophie stood up, disappointed. She laid a hand on the rock, peering over the moor for a tell-tale russet flash, but she couldn't see any sign of the fox. What had scared it? She bit her lip, and wondered why she couldn't just go for a little walk along the clifftop? What was the difference between walking here, and walking there?

Tentatively, she put a foot over the unseen boundary, and immediately felt odd. Her vision seemed to lurch, like when you miss your footing and step off a pavement. Something grumbled in her stomach and she felt her bile rise. Inexplicably, she felt dog-tired, as though she could just lie down right there and sleep. Latin words swam in front of her vision like the fish from her dream.

Sophie stepped back, and immediately the sensation passed, apart from a lingering headache. She rubbed her eyes, and suddenly couldn't quite remember what she was doing there. She turned around and headed back to the house, hoping that Andrew might be able to locate some paracetamol for her.

By the time dinner came round, Sophie was famished again. Andrew's cooking really was addictive. This time, it was a pork roast, with golden, glistening crackling, roast potatoes, parsnips and thick gravy. Sophie wolfed it down, Mandy watching her with an amused look. 'That'll teach you for skipping lunch.'

'We've not really spoken about Colin,' said Sophie between mouthfuls of food. 'I mean, for fuck's sake? He came into my flat. Carved that weird pentagram. You had no idea he was… well, nuts?'

Mandy watched her eat for a moment before speaking. 'Look, you know as well as I do that Colin was… unorthodox.'

'You can say that again!' remarked Sophie, laughing almost maniacally. 'He was a psychopath! He could have murdered me in my bed!'

'I'm sure it wouldn't have come to that,' said Mandy quietly.

Sophie put her knife and fork on the plate. 'Don't worry. I'm not going to complain to HR or anything, get anyone in trouble.' She finished her wine. 'Do we even have an HR department?'

'I believe,' said Mandy slowly, 'that Colin was honestly operating out of some kind of very misplaced good intentions. I have no idea what was going on in his head, but I think in some mixed-up way he was trying to protect you.'

'Protect me from what?' said Sophie. 'And God knows what would have happened if he'd had it in for me.' Something suddenly occurred to her. 'Bloody hell, I was supposed to be here with him this week, wasn't I? Talk about lucky escapes.'

'He's dead, Sophie,' said Mandy mildly.

'Thank God.' Sophie opened the second bottle of wine. She glanced at Mandy and realised she had to calm it down a little. She was right. Colin was dead. He might have been some kind of occult-obsessed madman, but he'd presumably worked with Mandy quite a while. Sophie decided she needed to be a little less shrill – though the wine was very nice and going down very quickly – and changed the subject. 'Why is this place called Macha House, anyway?' she said. 'Is it a Cornish name?'

'Irish,' said Andrew, putting dishes of steaming-hot spotted dick and custard in front of them. 'It's from an old tale. About a woman who had an amazing gift, and her husband could not keep it secret, as she asked. She was a witch, of sorts, I suppose. She could run as fast as the wind. And once word got out, men tried to take advantage of her gift, and she was forced to run a race, against the king's horses. She protested, saying she was with child, but the king insisted. Macha won the race, but her children – twins – were stillborn, right there. So she cursed the men who did this to her.'

Sophie said, 'Wow. So what's the connection with here? With the twins?'

Mandy leaned forward. 'The twins liked the story, that's all.'

Sophie sat back in her chair. 'When am I going to meet these mysterious twins, anyway?'

Mandy smiled. 'Actually, they'd like to see you now.'

37

Inside

Days in Withered Hill: 189

On the evening of the twenty-first of December, the villagers in Withered Hill begin to decorate their houses with holly and ivy, fir wreaths wound with wire oak twigs hung on their doors, tall white candles in their windows. In the square by the town hall, the village brass band takes up residence, and Mr Purcell, the butcher, has his boys carry around trays of fat sausages wrapped in bacon. Michael Ellison, the landlord of The Farmer and Devil, mans a wide trestle table where free mulled wine is handed out in little glass cups. Chestnuts crackle on flaming braziers and Thaddeus Obermann skips through the throng, a sprig of mistletoe suspended above his head on a piece of wire. Carol Mountjoy from the Post Office wears a short red dress, hemmed with white fur, throwing her arms around every man she passes and planting a huge kiss on his lips, cackling, 'You've been caught by Christmas Carol!'

Wrapped in big coats and scarves, Sophie and Catherine stand on the edge of the festivities, sipping their mulled wine. Sophie says, 'The band is playing *Last Christmas*.'

Catherine nods. 'Mmm. Here in Withered Hill we keep our own calendar, but we are not immune to the outside world. Yule has been celebrated on the twenty-first for as long as we can remember, but we are not above aligning ourselves with modern times. Thaddeus always says that Christians co-opted Yule to ensure the pagans they conquered would more readily celebrate Christmas; I suppose we're just co-opting Christmas back.'

As if summoned, Thaddeus pirouettes over to them and bends his lanky frame forward, lips puckered.

Catherine laughs and says, 'A kiss for our randy old librarian? Why not. It's Yule, after all.'

Thaddeus embraces Catherine and kisses her until she wriggles free. Then he turns to Sophie. 'And what of our Child of Promise? A kiss for Yule?'

'Child of Promise?' says Sophie, glancing at Catherine.

'That's you, my dear,' says Thaddeus with a smile. 'You came to us in spring, when life was new and budding. And in deepest midwinter you become our Sun Child, defeating the powers of darkness and ushering in the next spring, paving the way for Mother Nature's triumphant return.'

'Another festival, another starring role,' sighs Sophie. 'I will never leave here, will I?'

'I'm not sure why anyone would want to leave Withered Hill,' says Thaddeus. 'Now, a kiss was being discussed…?'

Catherine takes her arm and steers her away. 'Perhaps later, Thaddeus,' she says, then to Sophie, 'You *will* leave us. It is the way of things. But when you're ready.'

'I am ready!' insists Sophie as they weave through the revelry towards the mulled wine table. 'I've been here for six months. I know everything there is to know about my life out there. I have lived again all those moments. I remember them perfectly.'

'It isn't about remembering them,' says Catherine, handing her a glass of mulled wine. 'It's about learning from them.'

Sophie looks at her, tears in her eyes. 'I'm weary, Catherine. I'm tired. I don't know what I'm supposed to do. I just wish someone would turn to me and say, look, you'll never leave here. Then I could just resign myself to it and try to get on with some kind of life.'

'You will leave here. You have to leave here. It's the way of things.'

Sophie's hand insinuates itself into Catherine's, their fingers entwining. She murmurs, 'What if I don't want to leave?'

'It makes no difference,' replies Catherine, squeezing her hand. 'You will leave. You have a job to do out there.'

Sophie pulls away and watches the band for a while. Then she says, 'I think I'm going to go home.'

Catherine's brow crinkles. 'But it's Yule!'

'I'm just not feeling it,' says Sophie, sighing.

'But you'll come to mine tomorrow for the feast?' says Catherine. 'And early? I need help in the kitchen.'

'Of course,' says Sophie, forcing a smile. 'See you tomorrow.'

–

As she walks around her darkened cottage, Sophie feels as though every sense is heightened, every nerve ending raw, every pore on her flesh is a pool of fire. Her breathing is shallow and quick and she feels light-headed and giddy. It is not the three glasses of mulled wine, nor the Yule spirit. It is something else. Something she feels almost on the brink of, as though there is an epiphany looming, a breakthrough, a decision to be made, a boundary to be crossed.

She has been in Withered Hill for six months. She is repeatedly told that she has to leave, but is halted at every turn from doing so. There is no way out, and yet she cannot stay. It is driving her mad, she thinks. Actually, literally mad. It is unfair and frustrating and she feels as though she is being toyed with. What if this is all some kind of… experiment? What if it's like one of those reality TV shows that Sophie's social media posts were filled with? What if the people of Withered Hill are insane? What if, what if, what if?

What if Sophie stops playing the game? Stops being the subject of the experiment? Stops being at the whim of the insane? Sophie walks to a small pile of books that has accumulated in her living room and pulls one from the middle, the stack toppling over. She takes it to the window and flicks through it by the light of the Yule moon, finding the page she wants and reading softly out loud to herself.

'But I don't want to go among mad people,' Alice remarked.

'Oh, you can't help that,' said the Cat: 'we're all mad here. I'm mad. You're mad.'

'How do you know I'm mad?' said Alice.

'You must be,' said the Cat, 'or you wouldn't have come here.'

A sudden sob escapes her. She remembers the pressure of Catherine's hand in hers, their fingers knotted. Catherine telling her that she has to leave, that it's the way of things.

She doesn't want to leave. She can't wait to leave. She says out loud, her voice hoarse and strangled, 'I must be mad. Or I wouldn't have come here.'

In the kitchen, Sophie finds half a bottle of wine in the fridge. She pours it all into a pint glass and drinks it down, grimacing. That's what she does outside, isn't it? According to those social media posts. Drinks, drinks, drinks. If she's going to take up that life again, then she'd better get into practice. The wine immediately fuzzies her head, making her not warm like the mulled wine did, but cold and sharp. Focused. She knows what she has to do.

She has to leave.

But not by road, not by the woods; she has tried that. It's impossible. Withered Hill does not want her to leave. Not yet. It wants to keep her until it decides she is ready to leave.

No.

Sophie Wickham will decide when she leaves Withered Hill. And she is ready now. She cannot go through another month, cannot do another Withered Hill festival. Catherine has hinted at Faunalia, coming up in February, when the revelry and passions are high. Sophie thinks of Catherine's hand in hers, thinks of her lips, thinks of what it would feel like to put her hands on her body. She cannot bear it. She cannot conceive of making that amorphous, confused, unfocused love she feels for Catherine real, and then having it taken away, having to be told that she must leave, when Withered Hill decrees it.

Better to go now.

Sophie quickly goes upstairs before her resolve weakens, and in her bedroom writes a note to Catherine, quick and heartfelt. Then she goes to run herself a last bath.

The next couple of hours run through Sophie's consciousness like a fractured, badly edited film. There is a lot of chaos and confusion, there are a lot of people, there is noise and hurry and movement and she is sick, leaning over a bed and vomiting hard and long. There is darkness and light and crowds and loneliness and silence and sleep and eventually she opens her eyes and she is in a hard bed in a bare room, the dull December morning light filtering through threadbare brown curtains. A monitor on a wheeled metal frame beeps intermittently beside her, and there is a snaking clear tube dripping fluid into her arm.

Her wrists are bandaged.

She stares at them, trying to remember, when the door opens and in walks Catherine, wearing the same clothes she had on last night, her eyes red-rimmed, her face lined.

'Sophie,' she says.

'Catherine,' says Sophie, her eyes filling with tears. 'I'm sorry.'

Catherine comes forward and sits on the edge of the bed.

Sophie asks, 'Where am I?'

'Doctor Morris's surgery.'

'How did I get here?' The memories are coalescing in her mind, like pieces of a jigsaw coming together. She says, 'It was you. You found me.'

'I wanted to check on you. I was worried about the way you'd just left the Yule festivities. Turns out I was right to be.' Catherine puts a hand on Sophie's. 'Oh, Sophie. I wish you'd have talked to me.'

Sophie's face crinkles. 'I didn't know what to say. I didn't know what to do. I didn't know I was going to… going to do *that*, until I did.' She heaves a hard, wracking sob. 'I just want to leave. I don't want to leave. It's so fucking hard, Catherine. I've no idea what's going on.'

Catherine reaches into her coat pocket and pulls out the note that Sophie wrote to her. It is still sealed in the envelope with Catherine's name on it. She says, 'I haven't read this yet. Should I?'

Sophie thinks about the words she scrawled on the paper last night, and bites her lip. 'I don't think you should.'

Catherine nods and hands over the letter. She says softly, 'I think I know what it says, anyway.'

Sophie holds her stare. 'And what are you going to do about that?'

Catherine leans forward and puts a hand on Sophie's face. Then she kisses her, softly but firmly, on the lips. When she pulls away, she says in little more than a whisper, 'That. Just that. For now.'

Sophie stares at the letter in her hands. 'I've ruined Yule for you, haven't I?'

'Everyone is worried about you. Doctor Morris says you should be able to leave tonight. We got to you in time. You didn't lose a lot of blood.'

'*You* got to me in time,' says Sophie. 'You saved my life.'

Catherine looks at her. 'Yes, I did. And now you owe me, Sophie Wickham. And the favour I ask in return is that you never, ever do anything like this again. And that you concentrate on doing things the proper way, on waiting for the right moment to leave, and to do what's required of you. Do you promise me that?'

Sophie nods.

Catherine stands up. 'Right. Well, as I had no help in the kitchen this morning, I'm behind on the Yule feast.

I'll be serving it later. I'll come and get you about six. Take you home to bath and get changed.'

'What will everyone think of me?' says Sophie wretchedly.

'Nobody will mention it.' Catherine smiles kindly and touches Sophie's face again. 'You are our Child of Promise, Sophie. We love you. All of us.'

38

Inside

Days in Withered Hill: 362

The scars from Yule have healed, or at least the physical ones. In the bright June sunshine, Sophie often finds it hard to believe that she even did it.

The taste of Catherine's lips and the feel of her body yesterday are still imprinted on her memory, in the way the indentations of their bodies are imprinted on Sophie's mattress. Catherine is falling in love with her. She said so.

Sophie takes her mug of coffee and sits on a dining chair she's taken out to the front of the cottage, in the full morning sun. What is she supposed to do now?

Catherine had left early, to go home and shower and get ready for the school day. Sophie thinks of her, standing by the blackboard, teaching the children. The thought of not seeing Catherine again…

Sophie drains her coffee and stands up decisively, and heads towards the village centre.

Noah Jones, in shirtsleeves and leaning on a stick on the bench in the village square, looks at her curiously. 'What did you say?'

'I said, I want to stay. Here. In Withered Hill.'

Noah rubs his chin. 'But tha's tried to escape I don't know how many times.'

'I know I have,' says Sophie, keeping her voice steady. 'But now I want to stay.'

'But we just told you that you can start to make your bower,' says Noah, puzzled.

'Yes, I know,' says Sophie again. 'But now I want to stay.'

Noah frowns, as though thinking hard about this conundrum. Then he says, 'Well, you can't. It's not the way of things.'

Sophie pauses. Should she tell Noah about Catherine? Perhaps he'd be angry. Catherine had said it wasn't allowed. Not really.

'Can't you... can't you just go and ask Owd Hob?'

Noah laughs.

Sophie says, a little angrily, 'What's so funny?'

'Owd Hob,' he says, and slaps his knee. His laughter turns into a cough, and he spits a gobbet of phlegm on the stone flags.

Not for the first time, Sophie has the sense that she's not quite in step with everything that has been her life for the past year. It is as though she has said something ridiculous, asked Noah to petition Father Christmas or the Tooth Fairy to let her stay. As though Owd Hob is something to be discussed in the dead of night, when the moon is high and the bonfires are licking at the black sky, not to be talked about in the heat of the midday sun while the business of living goes on all around them.

'Forget it,' says Sophie, standing up. She looks at Noah. 'I suppose... I suppose I'm just meant to leave everything behind me, am I? Walk out of Withered Hill and… and just forget it all?'

Noah looks at her for a long time, with an infuriatingly unreadable expression. 'Oh, you're close, Sophie Wickham. So very close.'

'More riddles,' she says, and stalks away from Noah Jones.

Everyone keeps telling Sophie she's so close, but she still feels a million miles away. She decides to walk into the woods again, to see if inspiration comes. When she last communed there, she was disappointed with the results. Nothing concrete. Just a sense of a bargain, a trade, an accord. She had given of herself, as her part of the deal. She had let herself sink into the moss, let the crawling tendrils explore her. They had taken of her, but what had they given in return?

Sophie walks along the treeline, her hands touching the trunks of the trees. So many varieties here. Thaddeus told her that if you look hard enough, you will find every tree that ever grew in these isles in Withered Hill's woods. She has read up in the library on trees, looking for answers, and over the past year has got pretty good at identifying them.

'What's a bloody bower?' she says aloud, her voice echoing in the dead, quiet space between the trees, the sunlight fracturing on the dense foliage. She remembers Thaddeus reading the dictionary definitions to her. *One, a pleasant shady place under trees or climbing plants in a garden or wood. Two, a summer house or country cottage. Three, a woman's private room or bedroom.*

She has to make a bower. Those are the instructions. Which means, not the first definition. That already exists, it cannot be made by human hand. She could be said to be in a bower now, walking on the springy moss in the shadows of the trees. So, if it is a thing she must make, or build, then it must be either a summer house, or a woman's chamber.

They cannot expect her to build a house. And why would she make a chamber? Where? Here in the woods? For who?

Owd Hob needs a wife.

It sounds like the buzzing of insects, or the susurration of the breeze through the leaves. Barely there, like a lover's whisper. A memory surfaces, then is snatched away. Samhain.

You probably won't remember much of tonight. That's the way of Samhain.

Owd Hob needs a wife.

Sophie feels she is on the verge of catching it, of snatching it out of the air in front of her.

Owd Hob needs a wife. And what would that wife need? A bower. A wedding bower. In which to welcome her husband. A lady's chamber.

But who is to be his wife?

Sophie remembers the Beltane festival, the play the children put on. The farmer, whose daughter caught Owd Hob's eye. And his plan to appease Owd Hob by offering in her place the woman from the next village, the bad woman, who no one would miss at all.

Sophie turns and runs towards her cottage.

–

Sophie sits in her living room, the papers and printouts and documents spread all about her. In her hand, she holds the list she had made, of all the bad points belonging to who she had come to think of as Sophie Outside.

Drinks too much; Shallow; Treats people badly; Self-absorbed; Wasteful – clothes etc; Liar; Cheat.

It's been staring her in the face all this time. Now she understands. The farmer and the woman from the tavern in the children's play. Owd Hob takes the worst of humanity, takes those pieces off the board. *Owd Hob needs a wife...* that's just figurative. Metaphorical. She picks up a sheaf of papers. This person. Sophie Outside. The old Sophie. That's who Sophie must leave behind. That's who must be wedded to Owd Hob. She has been brought here to free herself from the past, so she can go back out into the world unburdened by all that baggage.

She knows what she must do.

Sophie runs into the village, dragging a black bin liner, and stops the first person she sees, Constable Parry, red-faced and sweating as he wheels his bike up the cobbled street. 'I need an axe!' she says. 'And a saw! A hammer, nails maybe. Or screws.'

Constable Parry pushes his helmet back off the hair plastered to his forehead. 'What you about, Sophie Wickham? Got a body to dispose of, or something?'

Sophie laughs. 'Yes, I suppose I have, in a way. The corpse of someone I no longer want to be.'

'There's the hardware shop,' he says. 'But I suppose anyone round here has those sort of tools.' He looks past Sophie and waves his hand. 'Hi! Peter! Sophie needs a little help.'

At the gate of Nut Nan Farm, Peter O'Keeffe hands the rough sack containing the tools over to Sophie. 'So you'll be leaving us, soon, then?'

'I think so!' she says happily.

'Where are you going to build your bower?'

Sophie points to the trees. 'Over there. This is where I came into the village, isn't it? A year ago. Do you remember? You were the first person to see me.'

'Aye, I remember,' says Peter.

Sophie holds the heavy bag with two hands and says, 'Thank you!' Then she turns and rushes towards the trees.

Sophie lays her tools out on the dry grass on the woodland floor and sits in front of them, communing with the trees, asking for their permission to take what she needs. She will build her bower and it will be of the trees, of all of them, a tribute to the woods, to Owd Hob and his people. Then she stands and puts a hand on the trunk of the nearest tree.

'Rowan,' she says. 'For luck.'

Then she takes up her saw.

'Applewood, for fertility. Birch, for happiness. Ash, for healing.'

With axe and saw, Sophie takes of the woods a limb from each tree, strong and supple, to help her build her bower.

She ventures deeper into the woods, identifying each tree, giving thanks to it for its sacrifice.

'Beech, for prediction of weather. Pear, for prediction of love. Plum, the oracle of dreams.'

And deeper yet to where the mighty oak grows, for stoutness of heart. And the yew, for protection against evil. And the ash, at which Sophie bows her head.

'For healing,' she says. 'Cousin of Yggdrasill, the world-ash, in which all the multitudes of forever are contained. I honour thee.'

It takes all afternoon, and the work is hot and hard and sweaty. The trees present themselves to her at every turn, even the ironwood, the tree of the wise women, and the lime, which is for justice. She takes from the poplar, of which the usurper god's cross was made, and the spruce, that can absorb illness and disease. She takes from the willow, the witch-tree; and the larch, for protection against bad magic. The pine gives her endurance, the elm yet further protection.

She thinks of Catherine when she takes the saw gently to the lone cherry tree she finds, the tree that belongs to the moon and is for love and passion. Finally, she stops at the fir, the last tree in the wood.

'Bearer of divine light,' Sophie whispers, 'I am the Child of Promise and I would have your help, more than the other trees in these woods. It is time for light to triumph over darkness.'

Sophie works until dusk, feverishly and without rest, until the last piece of wood has been used. She steps back, dirty and exhausted, her hands blistered and cut. But she has done it. She has made her bower.

There is a sound behind her, a slow clapping of hands. She turns to see Catherine. 'How long have you been there?'

Catherine shrugs. 'Half an hour, maybe.'

Sophie looks from her back to her handiwork. 'And what do you think?'

It sits against a wall of closely growing elm trees, forming its back, and from it arch the branches she has cut, the wooden bars of a cage. Sophie has fashioned a door, hinged with tightly bound vines, and it is sheltered with a roof of pine branches. It looks cosy, and inviting. It is a suitable bower.

'Well done,' says Catherine.

Sophie goes to get the black bin liner she brought with her from the cottage. In it is every piece of paper documenting Sophie Wickham's life outside Withered Hill. She reaches in, takes a handful, and tosses them into the bower.

'I consign the old Sophie to the bower, for the delectation of Owd Hob to do with as he will,' she says, grabbing more papers. 'The selfishness. The shallowness.'

Another shower of paper falls into the bower.

'Death. Lies.'

Paper falls like confetti.

'The waste. The self-obsession.'

She empties the remnants of the bag into the bower.

'All of it. All gone. All to remain here, in Withered Hill. Owd Hob needs a wife. Well, have her. You're welcome to her.' Sophie feels suddenly so exhausted, so empty, that she could sleep right there. She turns to Catherine. 'What now?'

'Now, we wait,' says Catherine. 'For word to reach the right ears. You've done good, Sophie. Now it's time to rest. Come with me. Back to my cottage. Let's make the most of the time we have left.'

39

Outside

Days to Withered Hill: 3

Andrew and Mandy led Sophie up the second flight of stairs to the double doors that opened into the top storey of Macha House. She was about to meet the mysterious twins at last. She suddenly realised she didn't even know if they were male or female, or how old they were, or even what their connection was to anything that was going on here.

'What should I do?' she whispered fiercely to Mandy as Andrew opened the doors. 'How should I speak to them?'

'Be respectful,' Mandy whispered back. 'But not awkward. Just be normal.'

The doors opened into a small, carpeted lobby, with four doors leading off it. Evidently, the top floor had been converted into a suite of rooms. Andrew led them to the far door, knocked sharply twice, then opened it and nodded for Mandy and Sophie to go through.

Beyond the door was a dining room, similarly furnished to the main one downstairs, but on a smaller scale. And sitting at the table were the twins.

They were women, impossibly old, in Sophie's estimation, thin and bird-like. Pale-faced, with white hair piled on their heads, both wearing cream gowns, their almost translucent hands cupping glasses of dark wine. They looked like ghosts, or vampires, thought Sophie crazily. They were both staring at her with identical pale blue eyes. She felt like she should perhaps curtsey, or bow.

Mandy cleared her throat and said, 'Grandmother. Great-Aunt. May I present to you Sophie Wickham.'

Sophie stared for a moment at Mandy. Grandmother? Why hadn't she mentioned this? And which one was her grandmother? The two women were not just identical in looks, but in mannerisms as well.

Simultaneously, they reached trembling hands for spectacles hanging around their necks and placed them on their thin, pointed noses to peer at Sophie.

'Come forward, dear,' said the woman on the left.

'Let's have a better look at you,' said the woman on the right. Even their voices were indistinguishable.

'I'm grateful for your hospitality,' said Sophie, searching for something to say. 'It's a very beautiful house you have.'

'Our hospitality is all down to dear Andrew, I'm afraid. We can't take any credit for that.'

'But thank you for your kind words about our home. We hope that you feel comfortable here, Sophie. And safe.'

'Yes, safe. That is very important, Sophie. That you feel safe.'

Sophie tried to brush away her frown. 'Yes, I do. Thank you. Very comfortable. And safe. Of course.'

The woman on the left patted the table with her fragile hand. 'Come. Sit. You too, Amanda.'

Andrew hurried ahead to pull out chairs for Sophie and Mandy at either side of the twins, and Sophie sat down. Andrew placed two glasses in front of them and poured red wine.

'So, you live in London, Sophie?'

'How do you find that?'

'To tell the truth, it's a little lonely sometimes,' said Sophie. 'A person can get… lost in London. All those people.'

The twin across the table from Sophie leaned forward a little. 'Ah, yes. Sometimes it is good to be hidden though. From prying eyes.'

'I suppose…' replied Sophie.

'And the parties!' said the other twin. 'I remember the parties. I loved a good party.'

'We might be ancient, dear Sophie, but that does not mean we never had fun.' She looked at her twin. 'How old are we now, dear?'

'Are we ninety-nine? What do you say, Amanda?'

'You are just a hundred, Grandmother. We had a little celebration in April.'

'A hundred!' said the twin on the left. 'Fancy!'

'Did we get a telegram from the Queen?'

'I'm afraid not,' said Mandy. 'We thought it prudent not to mention to Her Majesty where you were living.'

'Quite wise,' nodded the twin on the right. She smiled at the other. 'But then, strictly speaking, I am considerably younger than you, dear.'

The other laughed, a brittle sound like tinkling glass. 'Quite right, dear, you are. Strictly speaking.'

Sophie glanced at Mandy, but the other woman didn't meet her eye. Instead she said, 'I think we have imposed on you ladies for long enough. And Sophie has an early start.'

'Quite right,' said the twin next to Mandy. 'We are due to retire, and Sophie must do all her work tomorrow morning.'

'It's very important work,' nodded the other. 'Keeps you safe, you see. And that helps us all. In the long run.'

Mandy smiled at Sophie and stood up, and Sophie followed suit.

Andrew said, 'I shall take Miss Scott and Miss Wickham down and then I shall be up to clear away the dinner things and prepare your chambers.'

'Very good, Andrew,' said the twin on the left. Then they began to converse in hushed murmurs, oblivious to the presence of anyone else.

Andrew led them both away; the audience was over.

'I'm going to read in the drawing room, if you want to join me?' said Mandy. 'We could get Andrew to fetch us a bottle of something.'

'She's your grandmother?' said Sophie, taken aback. 'What is this? Some kind of family business?'

'After a fashion, I suppose,' replied Mandy. 'How about that drink?'

'I might go for a run first, while it's still light,' said Sophie.

'Good idea to keep your fitness up,' agreed Mandy. She looked at Sophie for a moment, then said, 'I like you. I hope all this works out.'

Sophie frowned. 'Works out? The job, you mean?'

Mandy smiled tightly. 'Yes. The job. That's right. You go for your run. Come join me later if you like.'

Sophie went to her room to get changed, puzzling over what Mandy had said. But… she liked her. Sophie hadn't thought much of Mandy during the months she had worked for her, but that was because she hadn't got to know her. She decided she quite liked Mandy too, even if she didn't really understand what they were doing here, nor why there was all the secrecy.

She got into her running gear and did some stretches at the window, the sky paling and clouds gathering right out at sea, on the horizon. At the edge of her vision, she caught a movement, right near the first stone where she was yesterday. Was her friend the fox back? She couldn't see anything now, but on the way through to the garden, she grabbed a couple of pieces of pork from the leftovers bin and wrapped them in kitchen paper, then let herself out of the house.

Sophie jogged to the first stone and unwrapped the meat, placing it carefully in the grass. She scanned the moors but couldn't see anything other than rocks and grass and scrubland. She set off at a run for the next stone, thinking about the twins. A hundred years old, Mandy had said. She wondered at their lives, wished she'd asked more questions. Had they always lived at Macha House? Were their husbands long dead? What of other family? And what was their connection to the company? They seemed to know of her work, and why she was here. And that curious comment, about what she was doing, inputting random, meaningless data… keeping her safe? Keeping them all safe?

She wondered what Tom was thinking, that she hadn't been in touch. Wondered if he'd tried to call her, the phone going straight to her answering message. She could try the telephone in the study again, she supposed. When Andrew had caught her yesterday, she had felt as though she was doing something wrong, but Mandy hadn't actually said she couldn't make a phone call? She could continue the fallacy she'd spun to Andrew, say that her gran was ill, that she needed to check up on her.

Sophie supposed that it was some kind of security issue. Perhaps Mandy would insist on being in on the call? In which case, Sophie would have to deliver some confused, garbled, coded message to Tom and he'd think she was even madder than he probably already thought she was.

London already seemed so far away and so long ago, even though it was only her second day at Macha House. All that business with Colin, the pentagram carved into her floor... it felt like some kind of weird dream. The fact he'd cut himself a key to her flat, sneaked in there while she was out with Tom, and then killed himself, photos of her spread all about him... Sophie shuddered. She'd barely had time to come to terms with all that before being whisked off to the wilds of Cornwall for top-secret work in a rambling old house occupied by two ancient twins she wasn't actually sure weren't ghosts.

Sophie slowed to a trot as she came around the wall of the garden, finishing her first circuit. She'd had too much food and too much wine and she had a stitch in her side. Maybe just one more circuit, and she'd call it a night. The sun was sinking in the west and the temperature was dropping slightly already. It was going to be much cooler tomorrow, she thought. At the bottom of the cliff beside her, she could hear the waves crashing on the rocks. It was actually quite beautiful here. In a raw, wild way. Not the sort of place Sophie would choose to go on holiday – if she had a choice, if anyone actually wanted to take her on holiday, she'd be off somewhere hot, sun-kissed, azure seas and golden beaches, cocktails and yachts and...

She'd be back in London soon. She could meet Tom. Maybe she could float the idea of a holiday? Was it too soon? But she knew she was falling for him, hard. She knew it, just knew it. He was handsome and funny and a fabulous lover and… well, he obviously wasn't short of money.

On balance, the day he'd walked into her with his cup of coffee was shaping up to be the best day of her life for a long, long time.

The thought of meeting Tom again spurred her on to do one more lap of the stones. He kept himself in shape, she should do the same. Especially if she was going to be getting into a bikini in front of him. It was June already, might be hard to book something for summer now, but there was always later in the year? Maybe the Canary Islands? It was hot there up until October, even later, wasn't it?

As she headed towards the first stone, Sophie noticed something sitting on top of it. Her first thought was that the fox had been at the meat she'd left out, somehow jumped up and put it, or the paper it was wrapped in, up on the stone.

But as she got closer, she saw it was something else. Something familiar. She slowed and walked cautiously towards it, glancing back at the house to see if anyone was in the windows.

There it was. Sitting on the stone. The elf. The one that Tom had given her. She frowned and walked slowly up to it. Wait. It wasn't exactly the same. Its waistcoat was different, more orange. It had a slightly lopsided grin. It was like the other one, but different. And it was sitting on a piece of folded paper.

Sophie snatched up the elf and opened the sheet of paper. There were a handful of words written on there, in a hand she instantly recognised. She looked around, scanning the moors, but there was no one in sight. Then she looked back at the paper.

SOPHIE,
YOU ARE IN <u>DANGER</u>.
MEET ME HERE AT MIDNIGHT.
TELL NO ONE.
TOM.

40

Outside

Days to Withered Hill: 3

Sophie made her excuses to Mandy after her run, saying the day's work had tired her out, and went up to her room. She showered as best she could and dressed in outdoor clothes again, lying on her bed and watching the old analogue clock on her bedside table ticking painfully slowly towards midnight.

How had Tom found her? And what did he mean by YOU ARE IN DANGER? Was it something to do with the work they were doing? But if she'd signed the Official Secrets Act, which meant she was working for the government, whose side was Tom on? Could she trust him? Could she trust anyone?

For the last half-hour, Sophie stood at the window, trying to look out onto the black moor. She saw no movement near the standing stone. What if this was some kind of test? she suddenly thought. What if it was an elaborate ruse set up by Mandy to see if she would break her pledge not to talk about her work at Macha House?

YOU ARE IN DANGER.

She could not ignore that.

At five minutes to midnight, Sophie stole quietly out of her room and padded along the corridor to the staircase.

The house was dark and quiet. She inched down the steps, wincing at the slightest creak, and then went into the kitchen and slowly turned the key in the back door, the click sounding like a hammer blow in the silent kitchen. But then she was out, and in the garden, and pulling open the battered gate.

There was a figure standing by the stone. She walked over, her breath coming quick and shallow, her heart pounding against her ribcage. The moon suddenly emerged from behind a thin cloud, and there he was, dressed in walking trousers and boots and a fleece jacket.

Tom Gisburn.

'Sophie, thank God,' he said, taking his hands out of his pockets and opening them wide.

'Tom. What are you doing here? How did you find me? What did your note mean?'

'I'm here for you, Sophie. I need to get you away from here. I'll explain later.' He looked over her shoulder at the house. 'Come with me, now.'

'But how did you find me?'

Tom sighed. 'The soft toy I sent you?'

She frowned. 'What? It's got some kind of... what? Tracking device?'

'Nothing so technological. It's an... older method. Please, Sophie. I'll explain when we're away from here.'

He held out his hand and hesitantly she took it. Then she tried to step past the stone, and stopped.

He looked at her quizzically. 'What's the matter?'

'I can't,' she said.

'You can,' he said, and suddenly dragged her towards him.

Passing the stone felt like Sophie was leaving something of herself behind. Her legs buckled and she fell against him. Her heart was hammering and her stomach convulsed. She jerked away from Tom. Her head was swimming and she thought she was going to black out.

'I… can't… please…'

'Shit,' said Tom and pushed her bodily back over the boundary. Immediately, the sensation passed and her head cleared. She stared at the stone, and then along the dark grass to the next one.

'What have they had you doing? The work?' said Tom quickly.

'I'm… not supposed to say. I signed the Official Secrets Act.'

Tom barked a humourless laugh. 'Of course you did. It's bollocks, Sophie. Tell me. What have you been doing?'

'Just inputting data, like at work,' she said wretchedly.

'What sort of data?'

'Words. Random Latin words. Spanish. Numbers. All kinds of things.'

Tom swore under his breath. 'A binding. And a strong one. Shit.' He rubbed his beard and walked up and down outside the boundary. 'The words you copied… do they still exist? The papers?'

Sophie thought back to the last time she was in the study. She was sure the sheets of paper were on Mandy's side of the table. 'I think so,' she said uncertainly. 'Tom, what is going on here? You're scaring me.'

'You should be scared,' he said shortly. 'But don't worry. I'm here now. Everything's going to be all right.' He held out his hand again. 'Sophie, you're going to have to invite me in.'

She had a sudden flashback to the last time Tom was in her flat, when she'd asked him to join her in the pentagram that Colin had carved in her floorboards.

'*I'd like nothing more but… your circle of protection and all that. I very much doubt I'd be able to cross it unless you invited me.*'

She put her hand cautiously forward, then pulled it back. 'Tom, who are these people? Mandy? And who are *you*?'

'Sophie, I promise I'll explain,' he pleaded. 'But only if I can get you away from here. If I don't…'

She took a deep breath and grabbed hold of his hand. 'Tom Gisburn, I invite you over the threshold,' she said, feeling stupid at the melodramatic words even as confusion and terror was turning her insides to water.

Tom sighed, smiled and stepped past the stone.

Sophie led Tom through the dark house, not speaking until she had quietly closed the door to the study behind them. She turned to him and whispered, 'Seriously. Tell me what's happening before we go any further.'

'There's no time,' he hissed. 'You're in danger. We both are. We need to do this and get out of here now.'

'But who are these people? It's just a data-processing company. And a small one at that. What do they want from me?'

'It's not,' said Tom, casting around the table. 'That's just a front. They're much more sinister than that.' His eyes lit on a pile of papers. 'Is that it?'

She stood at his shoulder as he flicked through the handwritten sheets. 'Yes, that's it,' she said. 'What did you mean, a binding?'

Tom dug into his pocket. 'A way to keep you here. A way to control you. Like that pentagram on your floor. That guy who did it, he was part of this, too. All this so-called data inputting… it's a method of coercion, Sophie. A way of keeping you compliant. Fogging your brain.'

She remembered, suddenly, the last day in London, when Mandy had summoned her back to finish the work she'd failed to do. How she'd had a splitting headache, how something had happened, something she couldn't quite remember. But when she'd done it, when she'd inputted the random strings of words, she'd felt better. And she didn't care much what had upset her so in the streets of London.

'But how?' she whispered. 'That's like… mind control. It's like *magic*.'

Tom flicked the lighter he had taken from his pocket. 'This should sort it,' he said, and touched the yellow flame to the corner of the sheaf of paper.

'And Colin?' asked Sophie. 'He was part of this too? But what exactly *is* this?'

The papers had caught, and Tom placed them in the empty fireplace. He picked up a folder of more paper from the table. 'This must be tomorrow's work,' he murmured. 'As soon as we're away from here, I promise I'll tell you everything. Far away. And yes, Colin was part of this. They all are. That whole company, it's just a sham. A front for something bigger. Something that targets people like you.'

There was an audible click from behind them and a voice said, 'Is that why you killed Colin?'

Sophie turned round to see Mandy standing by the open door, holding a shotgun at her shoulder.

'Step away from him, Sophie,' said Mandy calmly.

'What are you going to do?' said Sophie. 'Fuck, Mandy, you've got a *gun*.'

'Yes, and I'm prepared to use it,' she replied. She twitched the barrel at Tom. 'Hands where I can see them.'

'Mandy,' pleaded Sophie. 'Put the gun down. I don't know what's going on here, but this is ridiculous.'

'You're in extreme danger, Sophie,' said Mandy. 'Move away from him.'

'That's what he said!' shouted Sophie. 'Who am I supposed to believe? Him? You? Andrew? The twins?'

She saw Tom gaping at her. 'Twins? You don't mean… they're *here*?'

Mandy raised the gun. 'I can't let you go now you know that.'

This was insane. This must be a dream. A nightmare. Sophie willed herself to wake up, to be back in her bed upstairs. No, to be back in her bed in her flat, this whole crazy episode nothing but a figment of her imagination. She looked desperately at Tom. 'You know about the twins?'

Mandy's finger tightened on the trigger. 'Oh, he knows about them. But he'll not be telling anyone else.'

Three things happened at once. The first was that Tom deftly flung the pile of papers in his hand towards Mandy, the sheets separating and engulfing her like a blizzard. Then Sophie felt herself roughly pushed to one side by Tom. Finally, there was an ear-splitting explosion, and Sophie realised that Mandy had fired the shotgun.

Tom cried out and tumbled backwards into the fireplace. His shoulder was red and bloody, his jacket ripped open. Mandy had shot him. Sophie rushed to him and dragged him away from the fire, the burning papers scattering around him.

'Run,' he whispered to her. 'Get out. The binding's undone. They can't control you any more. Can't stop you crossing the bounds. You can get away from here. I'll find you.'

Mandy was standing over them, holding the shotgun high, pointing it at them both. 'Nobody's going anywhere,' she said.

Tom's leg kicked out and caught Mandy's shin, and she listed to one side, the gun going off and bringing down a shower of plaster from the ceiling.

'Go!' said Tom fiercely.

Sophie ran for the door, just as Andrew, in his dressing gown, bleary eyed, appeared there. She put her head down and barrelled into him, knocking him off his feet. As she ran past him towards the kitchen, she glanced over her shoulder to see Tom heaving one of the heavy chairs through the study window with a huge crash.

Breathless but driven on by adrenaline, Sophie pounded through the garden and out of the gate, and ran straight for the standing stone. She took a deep breath and leapt over the boundary that had kept her inside before. Nothing happened. Tom had been right. Whatever she had been doing, all those things she had been typing, it had somehow affected her. Controlled her. Had that been happening to her since she'd joined the company? She thought back to weeks and weeks ago, when she'd met Colin outside the pub at Donna's leaving do. A chance encounter, she'd thought. Obviously not. From what Tom had said… they'd targeted her from the off. None of it was coincidence. She'd been brought to the company, given that mind-numbing job to do. Every step of the way, her every move had been choreographed and controlled, like a piece on a chessboard.

She turned to the left and ran through the darkness for the road that ran past Macha House. Did Tom have a car somewhere? He must have. She cast around, looking for it, but could see nothing. Thank God for Tom. Thank God he'd come and rescued her. It raised more questions than it answered, of course. How he knew so much about Mandy and the company and the things they'd been doing to her.

Questions for another time, when her and Tom were far from here, and safe.

When she got to the road, she turned back to the house. The study was glowing yellow and orange. The burning papers must have caught the carpets. She thought about all those old, dry books, fuelling the fire. Silhouetted against the flames, she saw two figures, one holding a gun. Then there was a flash, and a boom, as Mandy fired again. The other figure fell to the ground and she cried out. But Tom picked himself up and limped away, around the other side of the house, as Mandy stood there, presumably reloading. Then she followed him and was lost from sight too.

Sophie bit her lip, then set off at a run along the road, to circle back around the other side of the house. It had looked like Tom had taken a hit to the leg as well. He was in trouble. She had to help. As she turned on to the grassland again, the Volvo parked in front of the burning Macha House suddenly roared into life, the headlights blazing.

She stood frozen in the beams as the car swept up the track towards her, slowing as it approached. The window wound down and she saw Andrew, his face set like granite. In the rear seat were the twins, impassive, in matching dressing gowns.

'Where are you going?' Sophie asked.

'Somewhere safe. Get in.'

Sophie shook her head. 'No. I'm not going anywhere with you.'

Andrew looked at her. 'We tried to help. Just remember that. We tried to help you.'

Then he gunned the engine and the tyres span on the dirt, and he turned onto the road and sped to the east.

Sophie ran to the right side of the house. Flames were licking through the upstairs windows now and the roof was buckling. She hurtled towards the cliff edge, shouting Tom's name. But there was no sign of him or Mandy. Sophie crept cautiously to the edge of the cliff, and looked down. The waves were crashing on the rocks in the blackness. She looked around, peering across the moors. Nothing. Tom and Mandy were gone.

Unsure what to do, Sophie just turned back to Macha House, and watched it burn.

41

Outside

Days to Withered Hill: 2

As dawn broke and the sun painted the sea with pale glitter, Sophie sat on the dew-damp grass, her back against one of the half-buried stones. No one had come. Not the fire service, nor the police. Not anyone. Macha House was a smouldering, blackened shell, poking charred fingers at the blue sky. There had been no sign of either Tom or Mandy.

Sophie got to her feet and headed for the road. She had no phone, no cards, no money. She walked wearily east for an hour, until a white van passed her and pulled over. A man offered her a lift to Truro, and she sat in silence, staring out of the window, as he prattled on about second homes and unemployment and fishing rights.

At the train station, she went to the customer services desk and they let her make a reverse-charge call to her gran, who paid for a ticket to London.

When she eventually got to her flat, she sat numbly on the sofa for an hour, flicking between the TV news channels and scrolling through media sites on her laptop. There was nothing about a fire at a big house in Cornwall. She took a long, hot shower and slept straight through until Friday morning.

In the afternoon, Sophie went to her office, but it was all closed up, and the man who worked in the design agency on the ground floor said no one had been there all week. Her security pass didn't work and she rattled the doors of the dark, empty office, in which her phone was locked.

After visiting her bank to get new cards ordered and take a cash withdrawal – thankfully, Mandy had already authorised her promised bonus for Cornwall to be paid into her account – she bought a cheap phone and a pay-as-you-go SIM card and repeatedly called Tom's number, but it just clicked straight onto his voicemail every time.

She bought food and wine and sat in her flat, ignoring the food and drinking the wine, and didn't know what to do next. Apart from her gran, she didn't know anyone, not any more. She didn't have anywhere to turn.

So it was something of a relief when she woke up the next morning to the insistent buzzing of her intercom, and a visit from the police.

There was one officer, a fat, sweating northern man whose name Sophie instantly forgot. He said he was there about an incident in Cornwall. Sophie invited him in and sat quietly on the sofa while he said he was just there to ask her some questions, and she wasn't in any trouble.

'You look a bit rough, love, if you don't mind me saying,' said the policeman. 'How about I make us a nice cup of tea?'

Sophie sipped the strong, sweet tea the constable had made and was wretchedly grateful for it. She began to relax almost instantly, and he said gently, 'Can you tell me why you were at Macha House?'

'It was work,' said Sophie. 'Or so I thought. I work for a data-processing firm. Or I did. Mandy was my boss.'

'Amanda Scott,' said the constable, flipping through his notebook.

'Yes. She said we had to do some work away from the office. I had to sign the Official Secrets Act.'

The constable frowned. 'Do you have a copy of that?'

'No,' said Sophie, feeling stupid and foolish.

'That's all right,' said the policeman kindly. 'And you arrived... Wednesday?'

Sophie nodded.

He said, 'Can you tell me in your own words what happened on the night of the fire?'

Sophie began haltingly, then everything came out in a rush. Everything. Tom and the twins and the work and even her dream about being underwater. She didn't see the point of hiding anything. When she was finished, she felt spent, almost sleepy.

The constable made copious notes as she talked. She vaguely wondered why there was a uniformed officer dealing with this. Wasn't it something that CID should be handling? If not... Special Branch? If there was such a thing outside TV crime shows.

'It's our understanding, Miss Wickham, that Amanda Scott and her associates were part of a wider organisation. They've been on our radar for quite some time.'

'Organisation?' said Sophie. She felt dull-headed and not quite up to speed with the conversation. 'You mean... what? *Terrorists?*'

'After a fashion,' said the policeman. 'I can also tell you that we recovered Miss Scott's body from the sea yesterday morning. She appears to have fallen from the cliff and died instantly from the impact on the rocks.'

'And Tom?' asked Sophie. 'Tom Gisburn?'

The constable sighed. 'That's mainly why I'm here, Miss Wickham. Tom Gisburn was found on the cliffs perhaps a mile away from Macha House. He had suffered two gunshot wounds. I'm afraid he's dead.'

Sophie felt like she should cry, but she just felt empty and dead. And so, so tired.

The policeman hesitated and said, 'There's more, I'm afraid. Mr Gisburn has no surviving relatives. As the last person to see him alive, we're going to need you to identify the body.'

'Oh, God,' moaned Sophie.

'It'll be quick and I'll be with you all the time,' he reassured.

Sophie took a deep breath. 'All right. When?'

'I thought we could set off now…?'

Sophie nodded and gathered her phone and purse. As she locked up the flat, she said, 'I'm sorry, what was your name again? It went out of my head completely.'

'That's understandable,' said the policeman. He smiled. 'My name is Constable Parry.'

—

Sophie woke suddenly, a crick in her neck from where she had been leaning against the window. They were driving through countryside, between rolling hills and vast expanses of moorland. The sun shone down through the windscreen of the police car.

'Oh, God, I fell asleep,' she said, looking around. 'Where are we?'

'Not far now,' said Constable Parry. 'And it's under-standable, after what you've been through. You've had a very stressful time.'

'We're not in London,' remarked Sophie, feeling stupid. 'I thought we were going to see Tom's body.'

'We are. He was transferred to his hometown. I'm afraid we're going there. He's at a funeral parlour in readiness for burial.'

Sophie pinched the bridge of her nose. She felt like she was swimming through treacle. 'How long was I asleep?'

'About four hours.'

She gaped at him. 'Four hours? Where is this town?'

'Lancashire. And it's more of a village than a town.' Constable Parry looked at her and smiled. 'It's called Withered Hill.'

—

Half an hour later, the car slowed and stopped. Ahead of them, the road disappeared into thick woods, the trees overhanging and forming a dark arch. The road was blocked by a lorry, one wheel by the side of the road and three men working furiously at the axel. Sheep were milling around and more could be seen through the open door at the back of the truck.

'Oh dear, looks like Peter O'Keeffe has had some bad luck,' said Constable Parry, chuckling. 'I'd better go and help.'

'Are we here? Withered Hill?'

'Just the other side of the trees,' said Constable Parry. 'One tick.'

Sophie watched him talking to the farmers. She felt strange. Disembodied. She'd felt fine earlier. It was ever since... ever since she'd drunk that tea.

Constable Parry ambled back to the car and came round to her side, opening the door. 'Bad news, I'm afraid. Fuel leak and everything. Road's impassable. Even on foot. Too dangerous. I'm going to have to cordon the whole area off.'

'Constable Parry,' said Sophie in a small voice. 'Did you… Was there something in the tea?'

'Just a little thing to help you sleep.' He smiled. 'Quite standard in these situations.'

Sophie frowned. It didn't sound standard. 'Wait,' she said, putting her hand on his arm. 'Tom Gisburn told me that he had a huge family. You said he had no surviving relatives.' The thoughts felt thick and mushy in her head, and she found difficulty in assembling them properly.

'No relatives, big family,' agreed Constable Parry. 'One big happy family in Withered Hill.' Then he put his hand on her arm and helped her out of the car. He said, 'See that path?'

She shielded her face against the sun and looked where he was pointing, a couple of hundred yards into the moor, where there was indeed a thin track cutting into the dense trees.

'You go that way. There'll be someone waiting for you.'

'On my own?' she said.

'Aye,' the policeman said with a nod. He smiled. 'That's the way of things in Withered Hill.'

—

Sophie stood on the path at the start of the trees and looked back at the road. Constable Parry was laughing with the farmer, and they were both looking in her direction. The trees were thick and tightly packed, and the foliage overhead sucked up any sunlight from above. Sophie took a deep breath and pushed through the branches.

She walked for she didn't know how long, her feet keeping to the path as best she could. She cried out as a bramble raked her ankle, and a branch whipped her and scratched her arm.

How deep were these woods? She was sure this wasn't normal. This wasn't right. The policeman... he'd drugged her. He'd admitted it. That couldn't be right, could it?

Sophie glanced back over her shoulder, wondering if she should just go back the way she'd come. But then she realised she'd lost the path completely. Could she even find her way back outside? She'd been walking for a while. Surely she should be coming out soon. To the village. To Withered Hill.

A sudden movement to her right startled her. An animal. Or a bird. Then another, behind her. She panicked and moved forward again, picking up the pace. A trailing plant brushed her face and she screamed, and started to jog forward through the trees. Something scratched her leg, and tugged at her dress, ripping it. She didn't care. She just wanted to be out of here.

Her foot sank in something soft and muddy, sucking her shoe off. She didn't stop to get it. Panic rose within her and she began to run, properly run, flailing through the trees as the branches hit and picked at her, ripping her clothes. She felt her dress separate, and suddenly she wasn't wearing it. She'd lost her other shoe.

She was blind with terror, lost in the woods. Woody fingers clutched at her. She looked down and she was no longer wearing her underwear. She was naked, covered in scratches. She fell forward, into a muddy patch, and picked herself up. She no longer knew why she was here. No longer knew where she was going. In truth, she didn't even know who she was any more.

Just when she was about to give up, and lie down, and wait for it all to go away, the trees thinned. She could see light, and sunshine ahead. A patch of blue sky. She was almost out. Almost out of the woods. And into Withered Hill.

She cried out as a figure loomed up in front of her from behind a tree. It was wearing a mask. A rabbit, or a hare, covering its face. It carried a length of tree branch in one hand. It looked at her for a moment, head on one side, familiar eyes peering at her from behind the mask.

'Hello, Sophie,' said the figure, in a voice she recognised.

Then it lifted its hand and brought the branch down on her head, hard.

42

Inside

Sophie awakes to a headache, and when she touches her fingers to her temple, they come away slick with blood. She is lying on springy moss, her body naked and covered in scratches and mud, her hair tangled with vines and leaves.

She is in a cage.

Curved wooden branches form the bars of her prison, a roof of fir sprigs above her, the stout trunks of elms at her back. She is surrounded by pieces of paper, weathered and curling. She picks one up. It is a printout of one of her social media pages. She grabs another. It is the same. Her messages and posts, all printed out. Another is a copy of her birth certificate. A cutting about her graduating university. What the actual fuck?

And sitting on a stump, intently watching her, is the person in the rabbit mask.

Sophie moans, and sits up, and tries to cover her nakedness. She says, 'What have you done to me? Why am I here? Who are you?'

In reply, the other puts their hands to their mask and pulls it up, over their long, dark hair.

And Sophie Wickham's face stares back at her.

Sophie pulls off the mask and places it by the bloodstained branch on the ground beside her. She says, 'I'm sorry about that. I didn't intend to hit you so hard. Are you all right?'

'No, I'm not all right!' says the other Sophie, old Sophie, Sophie Outside. She's angry now. She's every right to be. They stare at each other for a long moment. 'Who are you? You're *me*.'

'Yes,' says Sophie. 'I am you. Or I will be.'

The other Sophie shakes the branches, but Sophie has made them stout and firm. 'Why am I in a cage?'

'It's not a cage. It's a bower.'

After finishing the bower, it was as though a veil had been lifted from Sophie's understanding. She spent the next two nights in Catherine's arms, and all slowly and gradually became clear.

Sophie Outside, Old Sophie, the Other Sophie… that wasn't *her*. It wasn't her former life. It was the life she was going to take up. She had not been remembering, all this time, all this past year, she had been learning. Readying herself. All that putting the old Sophie in the bower… it wasn't metaphorical. She had thought that it meant simply putting what she had thought of as her 'old life' in the bower, a ritual of some kind, a cleansing. She had thought that was how she achieved her freedom.

Now she knows different. She didn't have to make the bower to put an idea in there, a concept, the sum of what she considered to be her old self. She had to make the bower as a trap, a prison, a cage. For whom she now realises was not Old Sophie, her past, but Other Sophie, Outside Sophie. Her double. Her doppelgänger. The one who had to be enticed to Withered Hill, and captured, and held here.

Now Sophie understands. There can't be two of them out there. And *that* Sophie had to be brought to Withered Hill. So *this* Sophie could leave. An exchange. Taking one person out of the world, and replacing them with another. Now she understands Owd Hob's purpose.

'How can you be me?' says Sophie inside the bower.

'Do you remember when you were twelve?' says Sophie. 'When you saw someone in the graveyard?'

Sophie's eyes widen. 'The ghost!'

'Not a ghost. Owd Hob.'

'He cut my hair! I *knew* he was real.'

Sophie smiles. 'Oh, he is real. Very real. It's just that nobody believes in him any more. Or what he stands for. Apart from here in Withered Hill. And he chose you.'

'Chose me?' says Sophie in a small voice.

Sophie nods. 'Chose you. He goes out into the world, does Owd Hob, and he chooses people. Chooses those who won't be missed, who have led wasteful, pointless lives. Failures. Nobodies. Bad people.'

'I'm not a bad person.'

Sophie raises an eyebrow. 'But are you a good person, Sophie Wickham?'

She casts her eyes down. 'I... I don't know.' Then she looks up. 'But I still don't understand. I don't understand... *you*.'

'You remember having your hair cut by Owd Hob. He brought that hank of hair back here. And he planted it in the fertile earth. And through Faunalia, and Beltane, and Midsummer and Lammas, and Yule, and all the festivals, all the celebrations, he tended his tiny crop, and for many more Faunalias and Beltanes and Midsummers and

Lammases and Yules. Until, a year ago, his ministrations paid off, and his seedling grew and sprouted and blossomed.' Sophie stands and does a little twirl in her short summer dress. 'And here I am.'

Sophie stares at her incredulously. 'He... grew you? From my hair?'

'He did. And in that seed was the essence of you. Memories to be plucked from the air, like dandelion clocks on the breeze. Things only you and I know. Your deepest secrets. Passwords to your social media accounts. Those you have loved from afar and never spoken of. Those you have hated.' Sophie stares at her. 'Those you have killed.'

Sophie shakes her head. 'I haven't killed anyone!'

The other Sophie raises an eyebrow. 'Yes you did, Sophie. You tell yourself you didn't. And you almost believe it most of the time. But you did. You did kill her. Emily. Your sister.'

Sophie opens her mouth to protest, then feels the strength desert her body, and the memories she had always been so very good at locking away and not thinking about flood in to replace it.

–

Sophie was in the nursery again, the nursery that was kind of the living room of her flat but wasn't. She held the hare once more, swaddled in blankets, but this time it wasn't dead. It was wriggling and making a satisfied, chattering noise. Sophie was small, nine years old. She was holding the baby hare tightly, tapping its little nose with her finger. She was alone in the nursery, with the baby hare, and she took it to the windows, where sunlight was flooding in. There was a gravel drive below her, and she realised it was the house she'd lived in with her parents, the Cotswolds hills rolling off against the blue sky. Her mother and father were climbing into a car, and perhaps feeling the weight of Sophie's stare, her mother turned and peered up at the nursery window, and waved.

Sophie wasn't alone with the baby hare in the house, of course, not at nine. Her gran was somewhere, probably snoozing in front of the TV downstairs. When the car had driven away, Sophie turned away from the window and laid the hare down in the cot, moving the blankets away from it. Then she walked over to a rocking chair on which there was a cushion with a picture of Peter Rabbit her gran had crocheted when the new baby arrived. Sophie picked up the cushion, gripping its edges in both hands, and walked quietly back to the cot.

The baby hare gurgled and looked up at her. It seemed to be smiling. Sophie smiled back, and then put the cushion over its face, and held it there until it stopped struggling. Then she replaced the cushion on the chair and went downstairs to watch television beside her sleeping gran. She didn't look at the hare before she left the room, because just before the cushion went over its face, it changed from a hare to a baby girl, and Sophie didn't want to see that again.

–

Sophie says in a small voice, 'I… I was a child. I was too young to understand. I didn't know what I was doing.'

'You put a cushion over her face until she smothered to death. You knew that was wrong. And you honestly thought a couple of hours at a charity shop every other week would make up for that? Would balance out what you did?'

Sophie looks around fearfully. 'You're mad. You're insane. You need help.' She swallows. 'I need help. There was a policeman. He must be near here.' Then she stands up and rattles the branches and starts to shout *Help! Help! Help!* at the top of her lungs.

Sophie watches her a while, and then says, 'No one will come. Or at least, no one you want to. And Constable Parry will be of no aid to you. He brought you here, remember? He set you on the path to Withered Hill. He told you to walk through the woods. You came to us freely and of your own will, as is the way of things in Withered Hill.'

'Please,' whispers Sophie. 'Let me go. I won't say anything. I'll pretend this never happened. Please. Just let me go. I'll go back through the woods. You'll never see me again.'

Sophie smiles sadly. 'I'm afraid I can't do that.'

'I can give you money,' says Sophie. 'My gran… she lives in the house my parents owned. They left it to her to hold in trust for me. There's money, as well. I can get her to give me some. How much do you need?'

'That's interesting,' says Sophie. 'I'll bear that in mind when I go to visit your gran.'

Sophie's eyes widen. 'You stay away from her! You don't go near her, you understand!'

'Maybe I'll call her. Every week. Maybe I'll visit her. Take her on day trips. Go to visit your parents' grave. Our parents' grave. Our *sister's* grave.'

Sophie slumps to her knees. She looks up at Sophie imploringly. 'Why are you doing this?'

Sophie squats down until she is on the same level as her. 'Because you were given a life and you have squandered it. In the pursuit of things that don't matter.' She casts her arm around. 'You and people like you… you were gifted a whole world and yet you stamp on it and sully it and

329

ignore its wonders. Poison it and concern yourself with inane, pointless things.'

'I can be better,' whispers Sophie in the bower. 'I can be a better person. I've learned my lesson. I understand what you're saying. Let me go. I'll be better.'

Sophie stands up again. 'It's too late for that. Perhaps if you had been a better person after Owd Hob had chosen you, if you had changed, if you had led a better life, then you might not be here now. I would not have grown, would not have flourished. If you had been a good person, Sophie Wickham, I would not have taken root and blossomed. But your shallow, empty, wicked life nourished me. And here we are.'

Sophie grows angry in the bower again. She grips the branches and shakes them. 'Who are you to pass judgement on people? Who is this Owd Hob anyway? What right does he have?'

'He has the natural right conferred to him by the earth he tends, and which tends him,' hisses Sophie fiercely. 'He is both the land, and of the land. He is the air, and of the air. He is the water, and of the water. He is the fire, and of the fire.'

Sophie looks over her shoulder, towards Withered Hill, and frowns. Sophie glances back; the villagers are coming, walking slowly towards them.

'Help! Help!' shouts Sophie. 'I'm being held prisoner here! Help me!'

Sophie watches the other Sophie screaming for a while until she quiets and stares at her with a mixture of fear and anger. 'There's no point doing that. Nobody here will help you.'

'You're evil,' whispers Sophie hoarsely. 'Or mad.'

Sophie smiles back at her. She recites, "'But I don't want to go among mad people," Alice remarked. "Oh, you can't help that," said the Cat: "we're all mad here. I'm mad. You're mad." "How do you know I'm mad?" said Alice. "You must be," said the Cat, "or you wouldn't have come here.'"

Sophie grips the wooden bars of her cage. 'But I didn't come here. I was brought here. This is kidnapping. Somebody will come looking for me. They'll bring the police. You'll all be in trouble.'

Sophie smiles kindly. 'And who will come looking for you, Sophie? Who cares enough to do that? Who haven't you pushed away or cut ties with or burned bridges to? Who, Sophie? Who will come?'

Sophie feels as though the wind has been knocked out of her. 'Nobody,' she says, her voice barely audible. 'Nobody will come. Not now Tom's dead.'

'And you brought yourself here. You walked into the woods of your own volition. That is how it happens. How it always happens.'

Sophie looks at her with red-rimmed eyes. 'Always happens? You've done this before?'

'Since longer ago than anyone in Withered Hill can remember. And we'll do it again. You're not the only one, Sophie Wickham. You're not *special*. In fact, that's why you're here.' Sophie smiles. 'But you will be. Special. That's my job.'

'I don't understand any of this. What's going to happen to me?'

'I'm going to go and make you some food. I'll bring it to you at sundown. And then…'

'And then?'

Sophie smiles. 'Then it's your wedding night.'

She walks away from the bower, the other Sophie's screams ringing in her ears.

43

Inside

'Bitch!' roars Sophie at the retreating back of her... what? Doppelgänger? What is going on here? Is it the drugs that policeman gave her in her tea? Is it making her hallucinate? But the other Sophie is just... *her* in every way. The looks, the voice, the mannerisms. Make-up? Plastic surgery? Sophie puts her face in her scratched, dirty hands. But why? Why? Why would anyone do this?

Sophie pushes herself to her feet and examines the cage. Bower, the other Sophie called it. Doesn't that mean... a bedroom? She blanches and feels like she's going to throw up. *Then it's your wedding night.* What is going to happen to her? What are these people going to do? Who is this Owd Hob? She remembers the ghost, or whatever it was, in the graveyard. She'd forgotten it soon after, put it down to a dream. But suddenly the pale face looms large in her memory, the ragged black clothes, the wide, toothy grin. Then other memories surface, like bubbles popping on the surface of a pond.

She remembers after Emily died and the people came to see her. Police, she realises. She'd blanked that out, as well. But now it runs in her mind like a scratchy old film. When the lady took her to one side and said those things to her.

'Sophie? Do you know what happens to bad girls?'

'They go to prison?'

'Yes. Sometimes. Sometimes they go to prison. Sometimes, very bad girls who continue to be bad, though… well, something else happens to them.'

'Oh? What?'

'Sometimes, Owd Hob takes them for his wife.'

'I'm sure I would not want to be married to anyone called Owd Hob.'

'Those that do are generally better people on the other side of it.'

And then she remembers Niamh Glenister in the shop.

'You're going to become a better person too, Sophie Wickham.'

'I can see it in you. You've been marked out.'

'Perhaps we can spend time together, Sophie, when you are a better person.'

Niamh was one of them, too.

'Oh, God,' says Sophie, then doubles over, her stomach convulsing, and she lets loose a stream of hot, acidic vomit onto the scraps of paper scattered about the bower.

On her knees, clutching her stomach, she remembers something else. Driving home from university. Her father shouting at her, her mother trying to placate him. And then, just before the car ran off the road, something rising up in front of them, as though inflating, a dark, ragged shape, with a white face shining like the moon.

'Owd Hob,' says Sophie in a strangled voice.

The other Sophie was right. She'd been marked out from the moment she… She retches again, nothing left in her stomach. From the moment she killed Emily. Put a pillow over her face and held it there until she stopped wriggling. That policewoman… she was one of them. She might even be here, now. She had marked her out. And then Owd Hob had come to visit her, when she was a little older. And liked what he saw.

It's your wedding night.

Then he'd come to her again, when she was in the car, and caused her father to skid off the road, killing both her parents. Making sure she was alone. Making sure she would live a life where nobody would care. Where nobody would come to get her.

'No,' says Sophie through gritted teeth. She pulls hard on the curved bars of the bower, but they are steadfast. The light is failing in the woods, but above her, she can make out thick knotted vines where the bars meet the trees behind her. She reaches up and her fingertips brush the bonds, tantalisingly close but not close enough for her to get a grip on them. She needs some kind of tool. She casts around for a stone, falling to her knees and scrabbling through the scattered sheets of paper, her fingers sinking into the moss, looking for something hard and sharp and—

'Sophie.'

She jumps and scrabbles back to the trees behind her as a figure looms out of the encroaching dusk. Then she feels relief flood her body.

'Tom,' Sophie says numbly. Then: 'Tom! Oh my God. They told me you were dead. That was just to get me to come with them, I suppose.' She stands up and grips the bars and peers towards the clearing where the village must start. 'Quickly,' she hisses, 'you have to get me out of here. These people… they're insane.'

Tom frowns and inspects the bars of the bower, taking one in his strong hand and shaking it. 'This is a good job. Totally solid.'

'They're tied at the top,' says Sophie quickly. 'You can probably reach the knots. I don't know who they are. I think Mandy must have been involved in this. And those creepy twins. Quickly, Tom, she said she was coming back after dark.'

'She?'

Sophie takes a deep breath. 'There's another one. Of me. Sophie. I don't understand it. Please, Tom, I'll tell you when you get me out. We need to get away from here.'

Tom takes a step back, and smiles. Not the sort of comforting smile that Sophie is looking for. No, the sort of smile that turns her insides to water. 'Sophie, I'm not here to get you out.'

She shakes her head as the tears come and screws up her face. 'No. No, Tom. Please. Don't say that.'

'I was born in Withered Hill,' he says. 'My job is going out there and finding those that Owd Hob has chosen. And making sure that they come here.' He smiles again, and it's horrible. 'As for Mandy, and the twins… they were trying to *save* you, Sophie. Why did you think Colin carved that protective circle in your flat? I found his address in that book you had and went round to sort him out. I enjoyed that. And at Macha House… the binding was to keep you safe. All that work they had you doing, inputting those words and numbers. Very clever. A new tactic. Designed to make you fuzzy and hidden from us, more difficult to influence.' He smiles. 'It didn't work. You should never have invited me over the threshold. Anyway, Mandy won't be helping anyone again. Not after I threw her off the cliff.'

'But Tom,' she whispers. 'I thought… I thought you loved me.'

His smile melts away. 'As if anyone could love you, Sophie Wickham. Even *you* hates you.'

'Bastard,' she says, and spits the hot bile that clings to her mouth at him.

Tom shrugs.

'Am I really that bad?' she says desperately. 'There are *evil* people out there. Murderers. Criminals. Bankers. Politicians. Why don't you bring *them* here? Why do you pick on people like me who haven't done anything?'

Tom squats down on his haunches and looks at her. 'It's a fair question. It's because those sort of people… they're irredeemable. If they change their ways, it'll be noticed. It'll be questioned. It won't be believed. So we pick those who *could* have changed. Who could have been better. Those who, if they suddenly start living better lives, will please people around them, not make them suspicious.'

Something like hope flutters in Sophie's chest. 'People who could change? I can change, Tom. I can be a better person. Is that what this is about? Are you trying to scare me into changing? That's OK. I can change. I can do that. Please let me go. Tell the others. I agree to it.'

'I wish it were so simple, Sophie. But Owd Hob needs a wife. And he's chosen you.'

It's your wedding night.

'Please,' she says, the sobs coming hard and ragged. 'No. Please, Tom.'

'It's too late, Sophie. You had ample chances.'

'It's so unfair!' she screams suddenly. 'Why me? Why pick on me? There are worse people out there.'

'I know,' says Tom. 'Far worse people than you. But you were easy, Sophie. You had already disconnected from the world. You were already halfway to Withered Hill, though you didn't know it. And now… now you're here.

And the other Sophie will soon be gone from Withered Hill. And she'll salvage the ruins of your life, if she can, and make a better job of it than you ever did. And then...'

'And then?' says Sophie, hope rising.

'And then she will find someone else, someone who has spent their time poorly, someone who won't be missed. And if Owd Hob wills it, then they will come here, and the one that Owd Hob grows to take her place will go into the world. And eventually, by degrees, there will be enough of us out there.'

'Enough for what?'

'Ah. That we'll see, eh?'

'But what will happen to me?' The words stick in her throat.

'Owd Hob will come for you,' says Tom, his eyes shining with fervour. 'He will take you down, into the land beneath the land, and he will take his pleasure. And when he is done, he will eat you.'

Sophie retches.

'And when he has eaten you, he will fertilise the land, which he is, and which he is of. And in his garden he will plant a hank of hair, and from that he will grow another. From the badness outside, goodness will flourish. And I will go out in the world and I will bring her mirror-image here, and it will all begin again.'

'You are fucking insane.'

He will take his pleasure.

He will eat you.

'You can't hope to get away with this!' she screams.

'We get away with it all the time,' says Tom. His face is almost hidden in darkness now.

'People will miss me! Despite what you think. I can't just disappear. Someone will notice.'

'You aren't going to just disappear, Sophie.'

Behind him, there is a light, getting bigger. A pale beam from a flashlight, approaching.

'What do you mean?'

Then the light shines in her face, and behind it is Sophie, the other Sophie, carrying a basket on the crook of her elbow. 'I said I would bring you food.'

And then Sophie understands. Why she won't be missed. Why the hole she has left will be filled, and what with.

The other Sophie smiles at her and places the basket down in front of the bower. 'Reach through. Eat. You'll need your strength.'

'Fuck you,' says Sophie, reaching through and tipping over the basket, the bowl of stew splashing onto the moss, the bread rolling down the hillock.

'Suit yourself,' says the other Sophie, and steps back.

The other villagers have gathered in a wide semicircle around the bower, standing, watching, silently. Sophie screams at them from inside. 'You're mad! You're all mad! What is this place? Some kind of asylum? You're all suffering from a mass delusion!' Her voice cracks and she says, 'I can help. I can get you all help. Let me out. I can help.'

Then Tom starts to sing in a low baritone. Sophie strains in the bower to catch his words, but she can't. They are not meant for her ears to understand. One by one, the villagers take up the song, and start to sway in unison, their voices rising on the summer breeze.

'What are they singing?' asks Sophie in a whisper. 'I can't tell what they're saying.'

'You're not meant to,' says Sophie, standing up. 'It's a lament, for you. But a celebration. For Owd Hob.'

She turns away, and Sophie says, 'Wait! Where are you going?'

'My work is done,' says Sophie.

'But what is going to happen now?'

'They will sing, until it's time.'

'Time?'

Sophie shrugs. 'The nuptials.'

'They'll find you,' says Sophie suddenly, her voice a strangled whine. 'The twins. They survived the fire. They're still out there. They'll find you and they'll stop you. Stop all this.'

Sophie turns and walks back towards the singing villagers and takes her place with them. She doesn't know the song they sing, but it comes to her lips, and as her voice joins with those around her, the screams of Sophie Wickham rise in the darkening night.

44

Owd Hob

They know the song though they never learned it. They were born with it in them, like their hearts that beat and their lungs that breathe. And before they were here, before any like them were here, the song would be sung by squirrels or rabbits, whistled by birds and hummed by bees. And before even they were here, the song was sung by beasts long since disappeared from the earth. And before even them, when the world was silent, save for the roaring of water and the burning of fire and the rushing of air and the grinding of the earth's plates and joints as it settled into itself, the song was sung by the wind in the trees.

Sung for me, and those like me, precious few that we are now. And that is not the fault of the wind in the trees, nor the beasts that no longer roam, nor the bees, nor the birds, nor the squirrels, nor the rabbits. They knew me and my brothers and sisters for what we were, what we are: guardians, curators, servants, masters. It is the fault of the men that came, who worshipped us, then feared us, then forgot us, then spread like a pox across the face of the world, not living in harmony with it but shaping it and raping it and bending it to their will. Consuming it and choking it and taking, taking, taking and giving nothing back. It was the fault of the men that came and starved us of belief and gave their fickle fealty to other gods: progress and commerce and selfishness and war. It was the fault of the men that came and buried my brothers and sisters beneath concrete and steel, burned them in their forges and mills, suffocated us with their chimneys. And once they'd seen off my brothers and sisters, and forced those few of us into retreat and hiding, in places like this, they staked their claims on the land, and divided it up, and killed each other for ownership of that which can never be owned, for it is not for sale.

The song is sweeter for being sung by the few that still know it, and it draws me from my slumber, deep in my garden. It is time, then. It is time to meet my bride.

I rise through the loam and the dirt and dress myself as I go, clothing myself in the form that they expect of me, which they gave me themselves from the bottomless well of their fears and superstitions.

Owd Hob, they call me. And Owd Hob I shall be.

They called me many names before I struck the compact with them long, long ago that made the men and women of this place loyal to me, not to their new gods. The agreement that I would give them the bounty of this earth if they would honour me and procure for me a wife. For the plan, the grand design.

The rewilding.

Boggart, they called me. Goblin and elf. The Good Folk, the Fair Folk, the People of Peace, they called me and my kin. As they grew bolder, they tried to diminish me, make me a tale for naughty children. They called me fairy, pixie, sprite. Fallen angel, demon, devil, they said as they rallied 'neath the shadow of a wooden cross.

Owd Hob, they call me now, as they bring me up through the worms and the dead things.

I gather flesh and bone and eyes and hair and a tongue and teeth and fingernails and a cloak to cover me, all formed from the raw material of the earth. Which I am, and which I am of.

And their song gets louder and clearer and I know I am close now. I have crossed from my space to theirs, left behind the place they cannot go to and will never understand, which they name to try to clarify and classify: the underworld, or the crystal halls, or Annwen, or the Unseelie Court. No matter what they call it, I am here now.

I grow myself from the soil, blooming like a night-flower, and the villagers fall silent. I stand facing them, my back to the bower.

'I am here,' I say. 'It's good to walk among you again.'

And the people of Withered Hill, the people of Owd Hob, the people who are true and loyal to the earth, cheer and wave and whoop and call. And at the centre of them, facing me, is the girl I grew from a hank of hair.

I beckon her over and she approaches cautiously, unsure. I say, my voice gravel and mulch, 'Sophie. Do you remember we met at Samhain?'

She shakes her head.

'Of course not, Samhain being what it is.' I reach out my white, bony hand and cup her chin in my fingers. 'Be not afeared. I grew you. From yon one's hair. You have a great and manifest destiny, Sophie Wickham. You go to do Owd Hob's work, to make the world right again.' I raise my voice and cry, 'To make things as they should be!'

The villagers call and cheer again, until I hold up my hands.

'Now, 'tis time for your revels and feast. Go, rejoice, for we have excised a hole in the world, and are filling it with one of our own.' I smile, and drop my voice. 'For tonight is my wedding night, and I would be alone with my bride.'

They roar their approval, and disperse, and I wait until they are gone and the only sound is the whimpering from behind me, and finally, I turn to face her.

I raise a hand, and the stout timber bars of the bower warp and widen, to allow me inside. She is sitting with her legs drawn up to her, rocking, smeared in muck, her face black from the running of her eyes. She has soiled herself and stares almost blindly at me.

'Tha's a pretty little thing,' I say, and she moans. 'A merry little piece.'

I take off my hood, and show her my chalk-white face. I smile with my teeth made of stones, and poke out my tongue formed of wriggling worms.

'Please,' she begs. 'Please. Please. Please.'

'Little Sophie Wickham. A bad, bad girl.'

Her eyes snap to a focus, narrowing, staring at me. 'Why? Why me? Why are you doing this? There's worse than me.'

'There's better than you, too.'

'What are you?' she hisses. 'You're not a man.'

'That I'm not,' I say. I pull my cloak of night aside, to show her myself. She closes her eyes and rocks backwards and forwards again. 'Yet, I clothe myself in the look of a man. I have the hungers of a man.'

'They said… They said you're going to… to… ea-ea-eat me.'

'Eventually, aye.'

She convulses. 'This can't be happening.'

I squat in front of her, putting my hand on her knee, and she flinches but keeps her eyes tightly closed. 'That's what we said, my kin and I, as the earth was snatched from us and we were forced out of its cool green. And we retreated, and regrouped, and held these spaces, these last, true, loyal places, like Withered Hill. And we started to fight back.'

'Please don't hurt me,' she says in a small voice.

' 'Tis a great and honourable thing that awaits you.' I hold out my hand. 'Take it. Come with me. Let's not delay longer.'

'Where are you taking me?'

'To my chamber.'

She begins to weep, hard and hot and snotty, and shakes her head. I keep my hand there, and eventually she places hers in it.

'Good girl,' I say. And I start to sink back into the earth. I am the earth and I am of the earth, and now so is Sophie Wickham.

She begins to scream and scream and scream until she is silenced only by the cool soil filling her mouth.

45

Interlude

June 1944

She couldn't believe her luck. She was free. She ran out of the woods and into the night, her feet flying over the grass, skipping over rocks, adrenaline driving her on, away from Withered Hill.

The only problem was, while she had an idea where to go, she wasn't quite sure how to get there. But that was for later; all she had to do was put as much distance between herself and Withered Hill as she could.

She risked a look back; nobody was following her. They wouldn't be. Not when they were all concerned with fighting the fires in the woods from that crashed aircraft.

Margaret ran, then walked, then staggered until daylight. She was still on the Lancashire moors, with little around her apart from farms. In a high field full of grazing sheep, she found a tumbledown, abandoned building, just one room made of stone, its roof caved in. Probably a shepherd's hut or something. She could see no farm nearby, so she risked curling up in the corner and sleeping.

She woke at noon and washed in a stream, then walked the rest of the way into a small town with tall smoking chimneys that nestled in a valley to the south of where she had been walking. Even given she had slept rough, Margaret attracted the glances of the few men she saw on the streets, which she found agreeable. She asked one the way to the police station and presented herself there.

'Margaret Bailey,' she said when the desk sergeant asked her name. 'I'm terribly afraid I have no papers. Thing is, I'm in the Land Army. I was up north of here when that bomber crashed. Terrible affair. You'll have heard of it?'

'The B-24?' said the policeman. 'Awful business. All lost. You're not injured?'

'No, but my poor car was. Lost everything. Papers, money, the works.'

'You want me to get you back to your unit?'

'Thing is, I've just started a fortnight's leave. Was on my way home. Terrible bind.' She leaned forward a little so the policeman could look down her dress. 'Not quite sure how I'm going to get home now, until I can get fresh papers.'

–

It was dark by the time the police car turned off the narrow road and onto the dirt track that led to the big, foreboding house perched on the cliff, the black sea rolling wildly behind it.

'This'll be fine,' said Margaret, giving the policeman a peck on the cheek and letting herself out of the car.

She walked cautiously along the track as the car did a U-turn and headed back the way it came. There were lights on in the house, and she could see a figure silhouetted in one of the downstairs windows. A figure that gave her a frisson of recognition.

Taking a deep breath, Margaret strode boldly forward and pulled on the bell-rope. The door was immediately opened by a boy so tiny Margaret was surprised he could have reached the handle. He looked up at her and then frowned. 'Auntie... Margaret?'

There was a movement behind him, and the door was opened wider. A woman said, 'I am so sorry. He shouldn't even be up so late. But we lost his father, my brother, at Juno Beach and...' The woman's words died on her lips as she beheld her.

–

At first, Margaret, the other Margaret, wanted to call the police. Then she wanted to call a doctor to have Margaret committed. Then she wanted to call a priest to have her exorcised. Then, finally, when she had put her infant nephew to bed, she sat down and drank half a decanter of whisky and listened to what Margaret had to say.

It was only when she asked her if she had ever seen a raggedy man with a pale face that the other Margaret faltered and said, 'How do you know about that?'

'Tell me,' said Margaret.

'It was ten years ago,' she said in a whisper. 'I was at boarding school. It was just after some... some rather horrid business. A girl died. By her own hand, actually. Threw herself off the bell tower.'

Margaret said nothing, just waited for her to go on.

'I mean, nobody liked her, to be honest. She was a sly little thing. It wasn't my fault. Nobody liked her. It wasn't just me who said those things to her.'

'And then he came to you...?'

The other Margaret blanched and nodded. 'In the grounds one night. I shouldn't have been there. I was

348

meeting a boy from the village but he didn't show up. But *he* did. Old Pale Face. Said all kinds of funny stuff to me. Then grabbed my hair and cut a chunk of it off! I was mortified. The girls all laughed at me the next morning when I told them.'

'That was Owd Hob,' said Margaret. 'This is what he does.' She lowered her voice. 'He has chosen you. To be his wife.'

The other Margaret's hand flew to her mouth. 'His *wife*?'

'It won't happen now,' said Margaret. 'Not now I'm here, I don't think.'

'Tell me. Tell me everything.'

So Margaret did. She told her that she had walked into Withered Hill one day, naked and alone and shorn of any sense of herself, any sense of anything. And gradually, she had learned about what she thought was her life outside, which she would be allowed to return to, when she had worked out how to do that.

'But it wasn't my life before,' said Margaret. 'It was your life. And you were to be taken to Withered Hill, and I was to replace you, and live your life differently. Better.'

The other Margaret snorted. 'Better? Says who?'

'Says Owd Hob, and the people who live on his land.' Margaret shrugged. 'It is a different morality they live by. Older rules. Ancient laws. They think their way is better.' She gazed off into the middle distance. 'In many ways, it is. They want us to reconnect more with the land that gives us life. Be less concerned with fripperies. Act more as custodians of the Earth, rather than a pox upon it.'

'This is what they told you?'

'No,' mused Margaret. 'It's what I worked out. When I suddenly knew what I had to do, to make the bower.'

'Bower? Like a chamber?'

'Exactly like that,' Margaret nodded. 'For you. And something terrible would have happened to you. I don't know what, exactly, but there was this... presence. I don't know what. In the woods. And foul, horrible things would happen.'

'But it's wrong,' said the other Margaret. 'It's positively *vile*.' She frowned. 'But who are you, exactly? Why are you the spit of me?'

'He grew me,' said Margaret. 'From your hair. To take your place. Grew me in the... the manure of his last wife. After he ate her.' She shook her head. 'Don't ask me how I know. I just do. When I got away from their influence, it all seemed to... rush into my head.'

The other Margaret pulled a face. 'I feel sick.'

'Me too,' said Margaret thoughtfully. 'It's wrong. I realised it was wrong when that plane crashed into the woods. It sort of... lifted a veil. Pushed the scales from my eyes.' She looked at the other Margaret. 'It broke the spell, when the real world came crashing in.'

'But you're here now. It's not going to happen, is it? You'll protect me?'

'No,' said Margaret. 'But there are others. Like you. Like me. And there will be more. Owd Hob will be choosing them even as we speak.'

'What a horrible rotter,' said the other Margaret. 'He's evil. More evil than Adolf. Those poor girls.'

'Perhaps...' said Margaret. 'Perhaps... we could do something.'

'About these other girls?'

'Yes, about all the other girls that are to come.'

'You're saying we could save them...' said the other Margaret.

'Yes,' said Margaret. 'That's exactly what I'm saying. We could save them. Together.' She looked around the house. 'Your parents died in a boating accident. You live here alone. With your nephew.'

'Poor Andrew, yes. Now that his father has died.'

'Perhaps I could keep you company, then.' Margaret smiled. 'We could be sisters.'

'Twins,' said the other Margaret. 'Funny, that. Did you know I'm a Gemini? Do you know astrology? What sign are you?'

'Gemini,' said Margaret. 'Of course.'

46

Inside/Outside

The day after, the bower is empty. No one talks of it, and the men quietly take it down and build a small bonfire of the wood. Sophie feels strange, almost anticlimactic. She goes to visit Catherine.

'So, what now?'

'We leave,' says Catherine. 'I drive you to London.'

'When?'

'Soon. After the Arrival.'

–

The Arrival happens two days later. Everyone seems to know about it without talking about it. Even Sophie knows that it is a special day, and she puts on her nicest sundress and takes her hare mask and walks down towards Nut Nan Farm, where Peter O'Keeffe is mending fences in the sunshine.

'You'll be off soon,' he grunts.

'Yes. Tomorrow, I should think.'

'Shame we never tumbled.'

'Perhaps for the best,' says Sophie. She smiles. 'There's always the next one.'

'Speaking of which,' says Peter. He turns as his daughter Meg runs from the farmhouse, dragging a big brown sack.

'Daddy, Daddy, Daddy!' she calls happily. 'It's time!'

Peter reaches into the bag and pulls out a weathered, leathery pig's head. A real one. Hollowed out. He puts it on top of his head, as though it's a motorbike helmet, and pushes it down. 'Aye,' he says, his voice muffled. 'It's time.'

Sophie looks towards the woods, and puts on her mask.

The woman emerges, naked and covered in earth, fresh from Owd Hob's garden. She is curvy and has red hair, her flesh pale. She looks from side to side, uncomprehending, then sees Peter and Meg and Sophie, and staggers towards them.

'Help,' she croaks as she approaches. 'Please. Help me. I don't know where I am. I don't know who I am.' She looks from one to the other, Peter and Meg in their pig heads, Sophie in her hare mask, trying to work out if this is normal, if this is real.

Sophie says nothing. She just points silently up towards Withered Hill.

—

Sophie feels like she has been born again, just like that day she first emerged into Withered Hill. Everything is new and wonderful and exciting and terrifying. She yelps at the sheer volume of the traffic on the motorway, the trucks towering over Catherine's little car. At the service station she piles sweets and treats into a basket, which Catherine quietly puts most of back on the shelf. She laughs delightedly at the hand dryer in the toilets.

'It's going to take some getting used to,' says Catherine. 'And that goes double for London. You're going to have to lie low for a bit, until you get a handle on things.'

'I can't believe I'm going to live in London!' exclaims Sophie, bouncing up and down in the passenger seat. She pauses. 'But Sophie never had any money. I don't have a job now. What will I do?'

'We've given you money,' says Catherine. 'In your bank account. There's a phone in your bag, with all the apps on it. And you'll have to get a job. When you're settled in.'

'A phone? So I can call you?'

Catherine takes her eyes off the road for a moment to glance at her. 'No. This is it, Sophie. We never speak or meet again after today.'

Sophie spends the rest of the journey in silence. When they drive through London, she looks in awe at the towering buildings, at the throngs of people, at the wide, sluggish Thames. Catherine directs the Mini through a warren of side streets until they stop outside a pub. Sophie says, 'Farewell drink?'

'Something like that.'

It is the only pub that Sophie has been in that is not The Farmer and Devil, but she has an inherited memory of pubs, of bars, of nightclubs. She follows Catherine to the bar and watches as she orders two gin and tonics. This will be her life now. Drinks in pubs where no one knows who you are.

'There's someone I want you to talk to,' says Catherine. She points at a table near the back of the pub, at which a woman is sitting alone, nursing a glass of wine.

'Is she from Withered Hill?'

'No. She's Kara. The girl Jamie was seeing. I messaged her and asked her to meet us here.'

Sophie frowns at her. 'Why?'

Catherine shrugs. 'Closure, I suppose. Come on.'

They sit down at the table and the woman, Kara, looks from one to the other. 'This is about Jamie, you said? Have you heard from him? How do you know him?'

'I was… seeing him,' says Sophie haltingly. 'Before you.'

Kara nods. 'Ah. The famous Sophie. He talked a lot about you.'

'I can imagine what he said,' says Catherine. 'Did he ever hit you?'

Kara blinks, not expecting the question. Her eyes narrow, and then she says, 'No. Well, he raised his hand to me once. Not long before he fucked off to wherever he's gone to. I kneed him in the bollocks and told him that if he ever even thought about that again I'd castrate him.' She frowns. 'What is this all about, anyway?'

'Would you like to know where he is?' says Sophie. She thinks about Jamie's rotten corpse hidden in the scarecrow in the field.

Kara shrugs. 'Not particularly. Dodged a bullet there, I think, with him fucking off.' She raises her glass. 'Here's to Jamie and wherever he's fucked off to, and when he gets there, I hope he keeps fucking off and then fucks off some more.'

Sophie and Catherine clink glasses with Kara, and when they've climbed back in the Mini, Sophie says, 'What was the point of that? To justify killing him at Beltane?'

'Maybe, a little,' says Catherine, putting the car into gear. 'But also to show you that Kara is happier without him. Sometimes when people go away, they leave a hole in the world. Other times, when they disappear, good energy rushes in to fill that void.'

'I still don't think he should have died,' says Sophie. 'But I suppose I understand why he had to. I think. He hasn't gone, has he? His energy is still there. But put to better use. I could never have been the one to do it, though.'

'And that's why you're here, and not still in Withered Hill.' Catherine turns into a wide street and parks the car outside the flat. Sophie's flat. 'You see both sides of the coin, now. You're better than me, and better than Sophie Wickham used to be. That's what you're doing here.'

Sophie looks past Catherine at the house and the window of her flat on the first floor.

Catherine says, 'Tom's been down and sanded that stupid pentagram off the floor.' She puts a hand on Sophie's. 'So. This is it.'

'Can't you come in? Can't you stay the night? Can't we be together just one more time?'

Catherine smiles sadly and shakes her head.

'But I still don't really understand what I'm supposed to do? Just be Sophie Wickham?'

'No,' says Catherine. 'You have to be Sophie Wickham, but better.'

Days since Withered Hill: 50

Sophie walks from the Tube station, sweating in the late August sun and from the crush of people on the train. She stops to buy a coffee and a salad, and heads towards the office. She feels her phone buzz in her pocket and takes it out as she walks, flicking through the messages. She has friends. Old ones, who were surprised to hear from her after so long, and who look at her quizzically sometimes. When she asks why, they just shrug and tell her she's changed. But for the better. New friends, too. From work. From the hobbies she's taken up. And a text from her gran, asking her if they're still going to the farmers' market on Sunday.

And while she was devastated to watch Catherine drive away from her flat for what she knew was the final time, there was happiness waiting for her as well. Her name is Louise, and she is a friend of a friend. They met on a long Sunday walk in the country that had been organised, ending up on a drizzly July afternoon in a pub with a thatched roof, where they drank Guinness and talked of books and the theatre and movies and music. Sophie was hungry for Louise's knowledge and taste, and for more as well, as the other woman's hand drifted to her knee and lingered there while they talked.

They have been seeing each other for some weeks now, and Sophie is unsure where it is going to lead, but for now she is happy in the things they do together, in the long nights they spend chatting, and in the comfort they find in each other's arms.

She's reconnected with Liz and Jill, and been in touch with Donna, who's coming over from Dubai next month. And she saw Tom Gisburn once, though she didn't speak to him. He was steering a tall, attractive black woman by the elbow into a bar in Mayfair. Another bride for Owd Hob. Niamh Glenister had been true to her word, and sought Sophie out. Now that she is a better person. They have become quite good friends.

Sophie had thought she'd feel alone in London, overwhelmed by its size and volume and mass. She doesn't. She feels part of something. And not just her growing group of friends.

A woman in her fifties wearing a light trench coat passes her on the pavement, and their eyes lock for a moment. Lock, and flash with recognition, though they have never met. And flash with something else, as well, something deep and primal and dark and free.

The woman is one of them. Like Sophie. Perhaps from Withered Hill, perhaps from one of the handful of other places like Withered Hill. And she is not the only one Sophie has met, passed by on the street, glanced at on the Tube, or in a bar. They recognise each other by the brief glow of their eyes, which speaks of the green land, which they are and which they are of.

Sophie has no idea how many of them there are. But she feels them, like a spiderweb spreading out across London. And their numbers are growing. Slowly, gradually, but growing.

Sophie glances back at the retreating figure of the woman, then looks up and around at London, at its monuments to excess and its temples to greed, its tarmac and steel and plastic that covers the earth, and the tired, pinched, beaten-down faces of the people who hurry by, no time for anyone but themselves. Living lives growing more and more insular, less connected, though they think they are quite the opposite. Everyone feels alone, everyone feels unmoored.

Everyone leaves, eventually.

But sometimes, some of them come back. And come back *better*.

Yes, their numbers are growing, the people like Sophie. And when there are enough of them...

When there are enough of them, what then?

Acknowledgments

When I told someone I was writing a horror novel, they looked at me curiously and said 'But don't you write rom-coms?' The answer to which was yes, I do. I've also written science fiction. And Victorian fantasy. I write comics, I'm a journalist. I just write, really. And though these things are all different genres and media, they have one thing in common: people. I write about people, whether those people are falling in love or falling into the abyss. *Withered Hill* is about people, like all good stories are.

I would like to thank my tenacious agent Laura Williams of Greene & Heaton for believing in this book, and Kit Nevile at Canelo for taking it on and making it better. Thanks to all the Canelo team, to the copy editor, Jade Craddock, to Sarah Whittaker, who designed the absolutely stunning cover, and to everyone who's worked hard to get this book out into the world. Thanks also to my family, especially my wife Claire, for unwavering support in the face of the ups and downs of being married to a writer, and finally thanks to you, for picking this up and reading it. I hope it gives you nightmares. Just wait until you see what comes next...

David Barnett